Computational Music Science

Series Editors

Guerino Mazzola
Moreno Andreatta

More information about this series at http://www.springer.com/series/8349

Guerino Mazzola · Maria Mannone
Yan Pang · Margaret O'Brien
Nathan Torunsky

All About Music

The Complete Ontology: Realities, Semiotics,
Communication, and Embodiment

 Springer

Guerino Mazzola
School of Music
University of Minnesota
Minneapolis, MN
USA

Maria Mannone
School of Music
University of Minnesota
Minneapolis, MN
USA

Yan Pang
School of Music
University of Minnesota
Minneapolis, MN
USA

Margaret O'Brien
School of Music
University of Minnesota
Minneapolis, MN
USA

Nathan Torunsky
School of Music
University of Minnesota
Minneapolis, MN
USA

ISSN 1868-0305   ISSN 1868-0313   (electronic)
Computational Music Science
ISBN 978-3-319-83714-7   ISBN 978-3-319-47334-5   (eBook)
DOI 10.1007/978-3-319-47334-5

Cover image designed by Maria Mannone

Printed on acid-free paper

This Springer imprint is published by Springer Nature
The registered company is Springer International Publishing AG
The registered company address is: Gewerbestrasse 11, 6330 Cham, Switzerland

Ohne Musik wäre das Leben ein Irrtum

(Without music life would be a mistake)

Collage by Maria Mannone

Friedrich Nietzsche (1844-1900)

Preface

The authors Maggie O'Brien, Yan Pang, Guerino Mazzola, Maria Mannone, and Nathan Torunsky. Photo 2016 by Yunqing Fei.

The idea for this book came from Maggie O'Brien and Nathan Torunsky, two very smart undergraduate students taking Guerino Mazzola's honors course "All About Music" at the School of Music of the University of Minnesota.

When Mazzola was introducing the students to this topic, they asked whether there was a textbook, and Mazzola had to apologize for not having any easy book that would cover all of the course materials. He rethought the issue, contacted Springer's editor Ronan Nugent and eventually signed a contract with Springer and organized a course at this university where the authors could write such a textbook, also in collaboration with PhD students Maria Mannone and Yan Pang, who had already written a Springer book *Cool Math for Hot Music* with Mazzola and therefore could contribute to this new project, too.

This book is the result of this exceptional course where all perspectives—those of undergraduate students, of PhD students, and of their instructor—could be merged to create a text, including many attractive figures and references, that would help students document the course materials.

In fact, this book is far more than just a textbook. We have paid much attention to writing in a style that would be accessible and interesting to anybody who wants to know what music is about in its most general understanding. In this sense, *All About Music* is for all who want to know about music but never could make it into this often ineffable but vital topic. The authors hope that this book in its unique architecture will open the world's sounds to those who would like to understand what music is all about.

We are pleased to acknowledge the strong support for writing such a demanding treatise from Springer's science editor Ronan Nugent.

Minneapolis, May 2016   Guerino Mazzola, Maria Mannone, Yan Pang,

Maggie O'Brien, Nathan Torunsky.

More relaxed authors after work: Nathan Torunsky, Yan Pang, Guerino Mazzola, Maria Mannone, and Maggie O'Brien. Photo 2016 by Yunqing Fei.

Contents

Part IV Communication

Part I

Introduction

1

General Introduction by Guerino Mazzola

Summary. This introduction gives a general orientation about the book's topic, its philosophy, and its collaborative authors.

$$-\ \oint\ -$$

"Without music, life would be a mistake." This famous saying of philosopher Friedrich Nietzsche is a provocative motivation to study music. But not only as a special field of knowledge or artistry, in fact Nietzsche addresses the full meaning of life.

In view of this energetic background, we have decided to offer a course at the University of Minnesota to freshmen and sophomore students that would give a broad introduction to the world of music. This is a fairly ambitious enterprise: *All About Music*, isn't it too much to tackle in a three-hour course for beginners?

To be clear: I have written thick and complicated books about computational music theory and music semiotics, and I have written composition, analysis, and performance software. I know music as a scholar, and I know it as a jazz performer and composer. Therefore I am aware of what it means to claim *All About Music*.

Putting Nietzsche's provocation and my scientific and artistic background together, I decided to take another approach to music, one that would touch its ontological depth, its ubiquitous presence in human existence, and at the same time its diversified and precise shape as a field of technologically enhanced knowledge and of embodied expression.

To realize this program, I started from my general ontological scheme developed in my study of music semiotics [40]. It is a fairly simple scheme that comprises four dimensions: realities, semiotics, communication, and embodiment. We shall describe it in the following chapter.

This scheme implies a fairly adequate definition of music, namely that *Music embodies meaningful communication and mediates physically between its emotional and symbolic layers.* This approach is a novelty and radically different

© Springer International Publishing AG 2016
G. Mazzola et al., *All About Music*, Computational Music Science,
DOI 10.1007/978-3-319-47334-5_1

from all those books written in the spirit of "Music for Dummies." Our book is not for dummies, it is written for all those who want to know about the existential depth of music as a way of thinking, feeling, acting, and loving.

Our approach is also not avoiding computational or other demanding aspects, but it does not presuppose higher education in maths, physics, neuroscience, or music notation. All required concepts and tools are introduced and developed when necessary. This book is about understanding, not about impressing laymen.

Last but not least, I have to be grateful to my students, Maria Mannone, Yan Pang, Nathan Torunsky, and Maggie O'Brien, who have contributed to the creation of this book with beautiful and smart efforts in texts and images. Only their engaged work has made it possible for me to face the challenge of communicating the interdisciplinary variety of musical perspectives to a broader public.

With these ideas in mind, I wish you a fascinating reading! Guerino Mazzola

Short biographies of the authors:

Guerino Mazzola qualified as a professor in mathematics and in computational science at the University of Zürich. He was visiting professor at the Ecole Normale Supérieure in Paris. Since 2007 he has been a professor at the School of Music, University of Minnesota. He developed a Mathematical Music Theory and software (Presto and Rubato for analysis, composition, and performance) and has published 24 books, 130 papers, 24 jazz CDs, a jazz video, and a classical sonata. He is the president of the Society for Mathematics and Computation in Music.

Maria Mannone has Master's degrees in theoretical physics, conducting, composition and piano (Italy), and in ATIAM at IRCAM-UPMC Paris VI Sorbonne. She studied at Accademia Musicale Chigiana di Siena. Her compositions have been performed at the Festival delle Orestiadi di Gibellina and by the Orchestra Sinfonica Siciliana. She is a Ph.D. student in Music Composition at the University of Minnesota, investigating musical gesture theory under the supervision of Guerino Mazzola, and collaborating with the theoretical physicist Mikhail Shifman. She is the author of *From Music to Image, from Image to Music*.

Yan Pang is a Ph.D. teaching assistant in the School of Music at the University of Minnesota. A sample of her varied publications includes the album *Glory Times* (contract songwriter and music director) by China Scientific & Cultural Audio Video Publishing Company; the score "Solis Ortus" (SunRiver Competition winner) by China People's Cultural Publishing Company; the paper "Scene of Sichuan Opera" (co-author with Mingzhu Song) by Sichuan People's Publishing House; and the book *Cool Math for Hot Music* (co-author with Guerino Mazzola and Maria Mannone) by Springer.

Maggie O'Brien is an undergraduate student at the University of Minnesota, Twin Cities, Class of 2019. She began her studies with the music department, but is now interested in communications and music as they apply to alternative healing practices. Her love for music spans many genres and types of performance, from clarinet recitals to hip-hop dance. She would like to thank her teachers and mentors Dr. John Zimmerman and Dr. Rena Kraut.

Nathan Torunsky is an undergraduate student in the Honors Program at the University of Minnesota. Motivated by his love of birds, Nathan was inspired by the flute solo from *The Bird* from Prokofiev's *Peter and the Wolf* and began flute lessons. Today, Nathan continues to play flute and piccolo. Despite his passion for music, Nathan's primary academic focus is in psychology with specific interests in the cognitive and brain sciences.

2

Ontology and Oniontology

Summary. This short chapter introduces the global architecture of ontology of music which this book is going to discuss in detail.

$$-\oint-$$

This book is about ontology of music, including three dimensions: realities, semiotics, and communication. It also includes the extension of ontology to the fourth dimension of embodiment. We call this extension "oniontology" for reasons that will become evident soon.

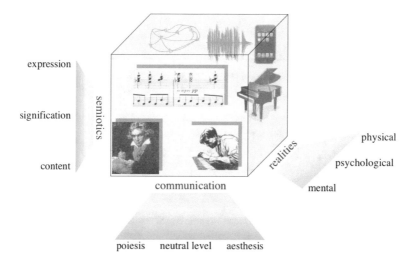

Fig. 2.1: The three-dimensional cube of musical ontology.

© Springer International Publishing AG 2016
G. Mazzola et al., *All About Music*, Computational Music Science,
DOI 10.1007/978-3-319-47334-5_2

2.1 Ontology: Where, Why, and How

Ontology is the science of being. We are therefore discussing the ways of being which are shared by music. As shown in Figure 2.1, we view musical being as spanned by three 'dimensions', i.e., fundamental ways of being. The first one is the dimension of realities. Music has a threefold articulated reality: physics, psychology, and mentality. Mentality means that music has a symbolic reality that it shares with mathematics. This answers the question of "where" music exists.

The second dimension, semiotics, specifies that musical being is also one of meaningful expression. Music is also an expressive entity. This answers the question of "why" music is so important: it creates meaningful expressions, the signs which point to contents.

The third dimension, communication, stresses the fact that music exists also as a shared being between a sender (usually the composer or musician), the message (typically the composition), and the receiver (the audience). Musical communication answers the question of "how" music exists.

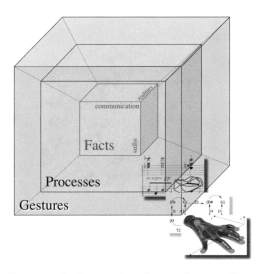

Fig. 2.2: The hypercube of musical oniontology.

2.2 Oniontology: Facts, Processes, and Gestures

Beyond the three dimensions of ontology, we have to be aware that music is not only a being that is built from facts and finished results, no, music is strongly also processual, creative, and living in the very making of sounds. Musical performance is a typical essence of music that lives, especially in the realm of improvisation, while being created. The fourth dimension, embodiment, deals with this aspect, it answers the question "how *come into* being?" It is articulated in three values: facts, processes, and gestures. This fourth dimension of *embodiment* gives the cube of the three ontological dimensions a threefold aspect: ontology of facts, of processes, and of gestures. This four-dimensional display can be visualized as a threefold imbrication of the ontological cube, and this, as shown in Figure 2.2, turns out to be a threefold layering, similar to an onion. This is the reason why we coined this structure "oniontology"—sounds funny, but it is an adequate terminology.

Part II

Realities

3

Physical Reality

Summary. To begin our discussion of the realities of music, we start with the most basic: the physical reality. It consists of the features of the world that we see in action every day. This includes the variety of ways in which music occurs and can be described, such as through sound waves, instruments, and human motion. There are different ways in which we can create and analyze music. We discuss Fourier analysis, frequency modulation, wavelets, and physical modeling as well as the hearing with the ear and brain.

$$-\ \oint\ -$$

3.1 Physical Sound Anatomy

Music, like the human body, functions as a system of parts. Therefore, we can approach the study of music as we would anatomy. The word *anatomy* comes from the Greek ανατέμνω (*anatemno*) and roughly means *to cut into parts*. Like a dissection, we can cut the 'body of sound' into parts. Fortunately, it is not necessary to kill music to study its sound anatomy! However, you will need some easy mathematics. Those who want to know more about the physics of music and its instruments, are referred to [16].

In this section, we will investigate sound waves (Section 3.1.1), and how to combine simple sinusoidal waves to build more complex natural sounds. Moreover, we will explain the difference between acoustical *frequency* and musical *pitch*.

Here is an example of the sinusoidal function, a mathematical tool key to the analysis of sound:

$$y(t) = A \sin(2\pi f t) \tag{3.1}$$

A sinusoidal function is produced from a regular rotation of an arrow of length A around the origin of the plane, as in Figure 3.1. The speed of such

© Springer International Publishing AG 2016
G. Mazzola et al., *All About Music*, Computational Music Science,
DOI 10.1007/978-3-319-47334-5_3

rotation is given by the *frequency* f, it tells us how often per second the arrow traverses a given position. The unit of frequency is Hertz (Hz). For example, if $f = 440\,\mathrm{Hz}$, the arrow traverses a given position 440 times per second. This is the frequency of the musical reference tone "chamber A" or "concert A". The vertical level of the arrow at time t is the function's value shown in Formula 3.1.

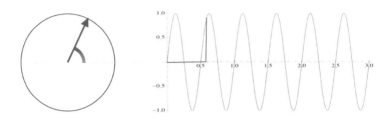

Fig. 3.1: A sinusoidal function from rotation.

A sinusoid is a mathematical object, not a sound, but is a formal tool that allows us to describe sounds. In order to better imitate natural sounds, we can mix sinusoidal functions: we can superpose waves (Section 3.1.2), we can put one sinusoidal wave inside the argument of the other (Section 3.1.3), or we can change the amplitude as starting and ending in zero (Section 3.1.4). All these techniques are used to build mathematical models that are approximations of what happens in nature (Section 3.1.5).

3.1.1 Acoustics

Sound is produced by a sound source. In music this can be an instrument such as the piano (see Figure 3.2). The strings of the piano vibrate and cause the air to expand with variable pressure. These changes in pressure occur at $343\,\mathrm{m/sec}$ in spheric surfaces (that is, the sound waves are spherically shaped). These pressure variances are then reflected off the parameters of the performance space and eventually reach the human ear.

A vibrating string of a piano or violin can be thought of as a system of small mass points that are connected by springs, see Figure 3.3. Upon displacing these masses (by the piano's hammer or the violin's bow) the springs' forces make the system vibrate. We shall come back to this model in section 3.1.5.

Sound waves are a result of differing air pressure. Regions of high density are called compressions and points of low density are called rarefactions (see Figure 3.4 where the air pressure is shown as spring deformation). These points correspond to the crests and troughs of the curve, respectively, in the sinusoidal function. We can distinguish the amplitude A as the distance between the highest (or lowest) point relative to the average pressure. The wave length λ is given by the distance between two consecutive peaks. The period P is the

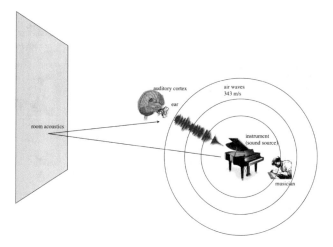

Fig. 3.2: Sound propagates from an instrument through air waves with the speed of 343 m/s.

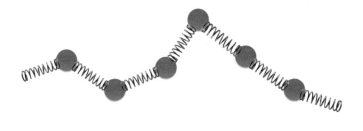

Fig. 3.3: A spring model of a vibrating string.

time interval after which the function is repeated, measured in seconds. How many periods in a second? This is precisely the frequency $\frac{1}{P} = f$. We have $\lambda = \frac{334\text{m}/\text{sec}}{f}$. Generally speaking, amplitude corresponds to volume (the higher the amplitude the louder the sound) while frequency corresponds roughly to pitch (the higher the frequency, the higher the pitch).

3.1.1.1 Standard Sound Representation

There is a standard description of a sound, see Figure 3.5. The sound is generated from a periodic wave and an envelope. Here, an *envelope* refers to the a shape in which a periodic wave is fitted. Therefore, we may focus on the wave, which together with the shape of the envelope is strongly responsible for the sound's color (the instrumental character).

Observe however that we cannot have music in a vacuum because there is no medium for the propagation of the acoustical wave. Without air, there

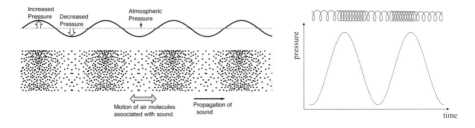

Fig. 3.4: Variation of air pressure in a sound wave.

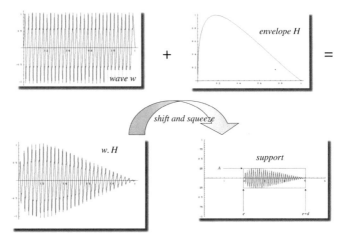

Fig. 3.5: Standard sound model: The sound is generated from a periodic wave and an envelope.

is no music. So, no music in the space between planet (even if the film music composers refuse to accept it). Light propagates in the vacuum; sound doesn't.

If we vibrate the middle point of a string (e.g. by plucking an instrument), we get the first path of oscillation as shown in the top of Figure 3.6. This is called the *fundamental* vibration of the string. If we keep the middle point fixed and vibrate the other two parts of the string, we get the second image of Figure 3.6, and so on. In this way we obtain the *normal modes* of oscillation. The fixed points are called *nodes*, while the points of maximum amplitude of motion are called *antinodes*. The second mode generates the octave of the fundamental; the third mode a musical fifth (such as C - F). In fact, that's the origin of the frequency ratio 2 : 1 for the octave, 3 : 2 for the fifth, 4 : 3 for the fourth, and so on, see also Figure 3.6. As you can see, the length of the vibrating string is inversely proportional to the frequency: the shorter the string, the higher is the frequency. This can be demonstrated by the equation:

$$\lambda_n = \frac{2L}{n}, \tag{3.2}$$

where L is the length of the vibrating string (fixed at the extremes), λ_n is the n-th wave length, and n is the index of the nth mode.

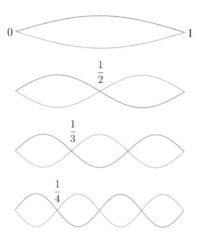

Fig. 3.6: Normal modes of a vibrating string.

What about the wind instruments? A bassoon player would be angry if we didn't let them know about the nature of his/her instrument. We should talk of it (b) as soon as possible. Here is the explanation: Air inside of a wind instrument can vibrate in the same way as a string. The same structure with nodes and anti-nodes is an air column, as it is in a vibrating string.

Waves in a string with fixed extremes are called stationary. They can be obtained as a superposition of two waves, one coming from left, and the other from right.

How is sound produced? The vibrating string provokes waves of rarefaction and compression of the air. Particles of air are periodically bouncing into and off of one another. In this way, the sound wave propagates. When the particles in a medium are oscillating along the propagation's direction, we talk about *longitudinal waves*. Sound waves are longitudinal, while other kinds of waves (e.g. electromagnetic) are transversal. The oscillation of particles from a vibrating string is transversal with respect to the direction of propagation of the wave, therefore they are transversal; see Figure 3.7.

When we have a real sound, we don't just have one mode, but in principle many or even an infinity of modes that are superposed.

Composers have also been inspired by the mathematical model of superimposed modes, and they tried to make music by recreating these sequences. An example of such *spectral music* is the piece *Partiels* by Gérard Grisey. We shall discuss this situation in Section 3.1.2.

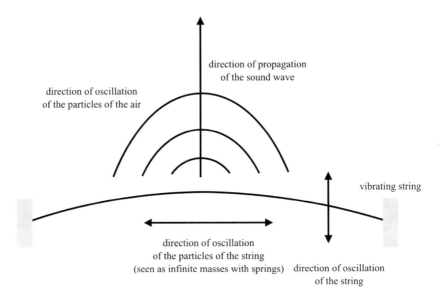

direction of propagation
of the sound wave

direction of oscillation
of the particles of the air

vibrating string

direction of oscillation
of the particles of the string
(seen as infinite masses with springs)   direction of oscillation
of the string

Fig. 3.7: Propagation of air waves generated by a vibrating string.

3.1.1.2 Pitch vs Frequency, Loudness vs Pressure

Whenever we play a sound with frequency f, we hear a pitch that is the *logarithm* of f. The pitch associated with f is defined as

$$Pitch(f) = \frac{1200}{\log_{10}(2)} \log_{10}(f) + C,$$

where C is a constant. Pitch is measured in units of *Cents*, Ct. To understand this formula, let us calculate the difference of $Pitch(f)$ and $Pitch(2f)$ one octave higher. This yields 1200 Ct. This means that the 12 semitone steps (such as $C \to C\sharp, C\sharp \to D, \ldots$, if they are set to be equal, the 12-tempered tuning) have 100 Ct each, i.e., one Ct is the hundreth of a semitone. Also this difference is the same for all f. This constant octave distance is visible on the keyboard of a piano: All octaves are equal in key distance. The piano realizes the above logarithmic formula. This is very fortunate. Imagine a piano keyboard, where the key distances were proportional to frequency!

Humans do in fact not perceive frequency directly, but the pitch that is associated with frequency. This is known as the Weber-Fechner law of (psycho)acoustics, because it explains the logarithmic perception of pitch.

By Weber-Fechner's law, we can establish a similar relation between the pressure amplitude A of a sound wave and loudness L:

$$L(A) = 20 \log_{10} \left(\frac{A}{A_0} \right) + C',$$

where C' is a constant. Loudness is expressed in dB (read "deciBel") and A_0 is the threshold of sound pressure amplitudes that can be heard by humans, equal to 2.10^{-5} N/m² (Newton per square meter).

3.1.2 Fourier

The physicist Joseph Fourier (Figure 3.8) had the idea to superimpose sinusoidal waves creating the so-called *Fourier series*. Earlier, we discussed the standard representation of a sound: wave plus envelope. The wave is a periodic function of time t, meaning that it is repeated after a time period P. Thus, $f(t+P) = f(t)$. This corresponds to a frequency $f = 1/P$. Fourier discovered that any such wave can be obtained as a (possibly infinite) sum of sinusoidal functions:

Fig. 3.8: Joseph Fourier (1768-1830).

$$w(t) = A_c + A_1 \sin(2\pi ft + \phi_1) + A_2 \sin(2\pi 2ft + \phi_2) + A_3 \sin(2\pi 3ft + \phi_3) + ...$$

The summand $A_n \sin(2\pi nft + \phi_n)$ is called the nth *partial* or *overtone*. A_n represents its amplitude, while ϕ_n represents the phase (the time shift of the partial). The summand for $n = 1$ is called the *fundamental* of the Fourier representation. In general the amplitudes A_n can also be a function of time. See Figure 3.9 for an approximation of the sawtooth function by sinusoidal functions.

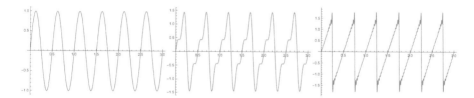

Fig. 3.9: An example of Fourier series approximating the sawtooth wave. From left: Fundamental, first three partials, first 20 partials.

Fourier's formula can be explained as a superposition of planetary movements, with overtones being moons, satellites, etc., see Figure 3.10.

Figure 3.11 represents an example of amplitude spectrum. The horizontal axis shows the frequencies, and the vertical axis shows their amplitude expressed in decibel level.

Fig. 3.10: Planetary movement as a metaphor for Fourier analysis. Think of the purple arrow and its revolution as a planet. The blue arrow then represents a moon revolving around the planet, while the green arrow represents a satellite orbiting the moon (which orbits the planet). This infinite continuation of this relationship is the fundamental concept of Fourier's series.

Attention: The decomposition of a wave into sinusoidal partials is not a divine law. In mathematics, it is shown that there is an infinity of other types of functions, not sinusoidal at all, which also enable such a representation with partials. The preference of sinusoidal waves is partly historical and partly due to the fact that the ear's cochlea also seems to perform a Fourier analysis (see Section 3.2.1).

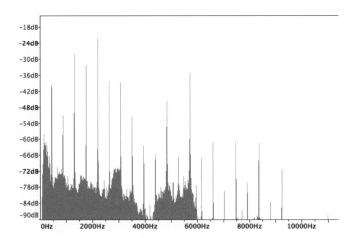

Fig. 3.11: An amplitude spectrum. The peaks correspond to the amplitude of the components in Fourier series, i.e., A_0, A_1, A_2, and so on.

3.1.3 Frequency Modulation

Fourier series are useful also as theoretical paradigm in mathematics, physics, and engineering, but they are difficult to handle in order to construct a technological device, for example. In principle, by opportunely superposing sinusoidal waves, it is possible to imitate the sound of different musical instruments. But, again, handling an infinite series is not possible. *That* would be an utopia![1]

A realistic solution to this problem was given by the construction of *Frequency Modulation* (FM) by John Chowning, in 1967, see Figure 3.12. The idea is to put one sinusoid inside the argument of another. This generates circularity, but it works because the circularity is not infinite: the last sinusoid will not have other sinusoids in its argument. Implementing this tool, you can play the sounds of an entire orchestra with a simple keyboard! Chowning sold his patent 1973 to Yamaha who produced the famous DX7 synthesizer, see Figure 3.13.

Fig. 3.12: John Chowning (1934-).

Let us delve into the mathematical details. In formulas for frequency modulation we can distinguish a *carrier*, that is the main sinusoidal function that contains the others, and a *modulator*, the sinusoid inside the argument of the main one. A modulator can be the carrier S for another function with its modulator, and so on, where A_1 and ϕ_1 are amplitude and phase of the carrier, A_2 and ϕ_2 amplitude and phase of the modulator, and A_3 and ϕ_3 amplitude and phase of the second modulator. Here is a typical FM function:

$$w(t) = A_c + A_1 \sin(2\pi f t + \phi_1 + A_2 \sin(2\pi 2 f t + \phi_2 + A_3 \sin(2\pi 3 f t + \phi_3))),$$

See Fig. 3.14 for examples of such FM systems.

In general, amplitudes A_i can be a non-constant function of time, similar to amplitudes in the Fourier formula. An example of frequency modulation with a non-constant amplitude is shown in Figure 3.14.

Frequency modulation is used not only in music, but in a wide variety of contexts: radio communication, EEG, radar, seismic prospecting. Another type of modulation inside a wave is the *amplitude modulation* (AM). In this case, the frequency is constant, while the amplitude is varying. It has been shown that

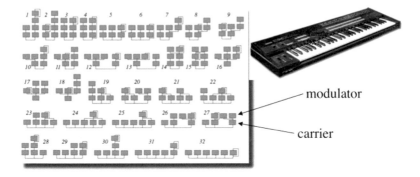

Fig. 3.13: Yamaha's DX7 algorithms of frequency modulation.

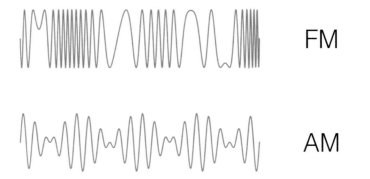

Fig. 3.14: Frequency versus amplitude modulation of a signal (FM versus AM). The formula of the chosen example of FM is $4\sin(t + 0.5\sin(0.1t))$, where $4\sin(t + ...)$ is the carrier, $0.5t\sin(0.1t)$ the modulator. The formula for the example of AM is $4\cos(0.1t)\sin(t)$, where the modulating amplitude is $4\cos(0.1t)$.

FM is more stable than AM with respect to signal noise. Figure 3.14 shows a situation of frequency modulation, as opposed to amplitude modulation.

3.1.4 Wavelets

Although FM is efficient, its functions are still of infinite duration, as opposed to real sounds. To solve this problem, a new tool (derived from the theory of signals used in music) was created: the *wavelet*. A wavelet is a wave whose amplitude starts from zero, increases, then decreases and returns to zero. See

---

[1] There was another Fourier, named Charles, who described a model of *utopistic society*.

Figure 3.15 for an intuitive explanation, using the concept of envelope (Figures 3.15, 3.16), and Figure 3.17 for examples of wavelets.

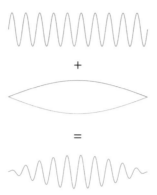

Fig. 3.15: An illustration of the concept of *envelope*. The top image, representation of a sinusoidal signal, is deformed accordingly to the shape of the middle image, modifying its envelope and giving the bottom image.

Wavelets are used to solve the signal-cutting problem. A wavelet is different from zero in a finite range. Wavelets are used in several fields, including seismography and heart-beat monitoring in electrocardiograms, ECG. Wavelets can be used to monitor the pulse of the Earth, and the pulse of the heart!

The idea comes from the theoretical physicist Denis Gabor in 1946, using a concept of quantum mechanics. In 1984, Jean Morlet [21] and his collaborators extended the concept, defining the continuous wavelet transformation that we will briefly describe soon.

Given a wavelet ψ, we can define an infinite set of ψ's deformations by the following formula:

$$\psi_{a,b}(t) = \frac{1}{\sqrt{a}} \psi\left(\frac{t-b}{a}\right),$$

where a and b are real numbers. The parameter a gives the width of the signal, while parameter b is related to the shift along horizontal time axis. See Figure 3.17 for a graphical representation.

It can be shown that any musical sound signal can be obtained by superimposing a number of deformed wavelets $\psi_{a,b}(t)$. This result corresponds to Fourier's theorem, but now, any signals, not only periodic waves are considered.

3.1.5 Physical Modeling

We have a set of primary colors, and by mixing them, we can obtain all colors we need to paint the images of nature. The same thing happens with mathematics.

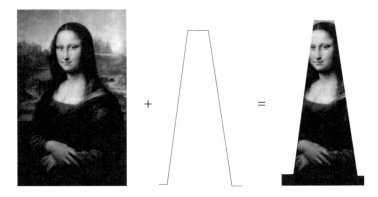

Fig. 3.16: A humor-illustration of the concept of *window* or envelope. If we have a signal, and we apply to it a window of a certain shape, given by a mathematical function, the result will be a signal-cut following the shape of the window.

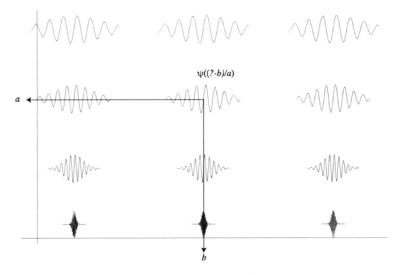

Fig. 3.17: Example of wavelets, with different values of parameters a (width) and b (shift). The question mark in the formula is a conventional notation for a not-yet assigned variable.

We use a set of 'primary functions' (e.g. sinusoidal functions) and by 'mixing' them we can make models that approximate the real sound waves in nature.

In this way, we can combine equations, mathematical 'colors', to create different sounds to change their *timbre*. With sound synthesis, scientists have alse been able to (almost) imitate the human voice! We talk about *physical mod-*

eling when we don't imitate the sound (like with Fourier, FM, and wavelets), but the sound source, the instrument.

With modern fast soft- and hardware, we can program physical systems of musical instruments. For example, using the mass-spring model described earlier, we can simulate a string's dynamics. The system "Cordis Anima" by Claude Cadoz and his collaborators (1989) works with the mass-spring model. Another strategy is to simulate air waves. This has been realized, for example, by Perry Cook. See Figure 3.18 for an example of such circuits and components for the simulation of the human voice. A third method has been implemented at the IRCAM in Paris. It constructs an instrument from its components in the sense of a Fourier synthesis by parts (the "Modalys" system). Physical modeling

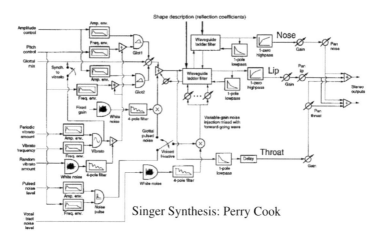

Fig. 3.18: Sound synthesis with physical modeling: an example of concatenation of circuits, and their role to simulate the human voice.

has been implemented in the music industry, e.g., by Yamaha in its synthesizer VL1 in 1994.

3.2 Hearing with Ear and Brain

Summary. As we have learned, sound is created by differences in pressure, typically vibrations traveling through air. However, our perception of sound is not directly related to differences in pressure. Our brain cannot receive this information directly. Instead, the human body uses a series of mechanisms to codify this information, provide the code to neuronal cells and send this neural code to the brain.

3.2.1 Hearing with the Ear

Your sense of hearing is the result of this translation and transmission of acoustical information. How does the body do this? The perception of sound begins in the ear, surely a not-so-shocking revelation. What might be less clear, however, is the variety of different components in the ear that help process sound. The ear has three main parts: the outer ear, the middle ear, and the inner ear. The outer ear is the portion of the ear that we are able to see, see Figure 3.19 for a visualization.

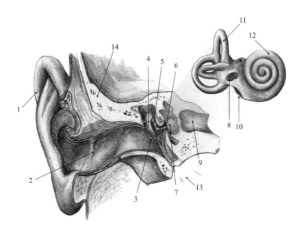

Fig. 3.19: Frontal section showing the parts of the human ear, with the detail of the osseous labyrinth. External ear: 1. pinna, 2. auditory canal; Middle ear: 3. tympanic membrane (ear drum), 4. malleus, 5. incus, 6. stapes; Inner ear: 7. (and 8.) oval window, 9. (and 10.) round window, osseous labyrinth, containing 11. semicircular canals, 12. cochlea; 13. Eustachian tube (in a plane anterior with respect to the section); 14. temporal bone. Drawing by Maria Mannone.

A sound wave's journey into the ear begins with the *pinna*, the cartilaginous formation responsible for filtering incoming sound. From there it enters the *concha*, the opening that leads to the ear canal. The *ear canal* is a short tube in the outer ear, and is the last part of the ear in which sound is transmitted through the air. It acts as a chamber in which sound can propagate through the air to the end of the outer ear, which is marked by the *tympanic membrane*, or the "eardrum." The eardrum is a thin, taut membrane that vibrates based on the differences in pressure present in the auditory canal.

The eardrum separates the outer ear from the middle ear. The vibration of the eardrum is passed onto three bones in the middle ear, referred to collectively as the *ossicles* or individually as the *malleus*, the *inca*, and the *stapes*. The ossicles are the smallest bones in the human body. Due to their size, they are highly sensitive to vibration, and can increase the power of these vibrations by a factor of 20!

The inner ear, called the *cochlea*, is filled with fluid. In order for the ossicles to create a pressure difference that will propagate well in the new, fluid medium they must concentrate the force of vibrations at the *oval window*, a bridge between middle ear and the cochlea directly connected to the stapes. In addition to the oval window, the cochlea has a *round window* which dampens the vibrations after having traveled through the cochlea. Without the round window, the residual vibration would cause us to hear everything twice, like an echo. This would inundate the brain with an excessive amount of auditory information.

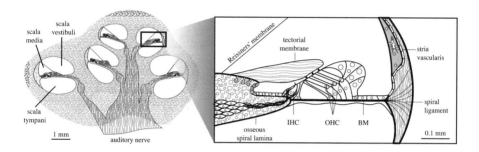

Fig. 3.20: Section of Cochlea. IHC: Inner Hair Cells, OHC: Outer Hair Cells, BM: Basilar Membrane.

"Knock Knock!" says the stapes. "Who's there?" asks the oval window. Sound! Finally, the vibrations have reached the cochlea, where the perception of auditory information begins to take place (see Figure 3.20). The cochlea is the sensory organ responsible for hearing. It is spiraled, like a snail's shell, but the total, unfurled length of the cochlea channel is 3 cm.

Inside this 'snail's shell' is the *basilar membrane*. The basilar membrane contains 20,000 neural hair cells responsible for coding vibrations in the fluid into a language of electrical signals. The basilar membrane is tonotopically organized, which means that specific locations are responsible for specific frequencies [4]. Neuroscientist George von Békésy discovered that high frequencies are perceived near the oval window whereas low frequencies are perceived at the apex of the cochlea. For this discovery, he was awarded the Nobel Prize in 1961.

3.2.2 Hearing with the Brain

Summary. Now that sound has been translated from pressure differences in the air to electrical impulses in the cochlea's hair cells, what happens next? In short, the body's equivalent of electrical wiring takes on the responsibility of getting that information to the brain, where it can be processed. This network is known as the *nervous system*, and it is based on electrical and chemical communication between neurons.

— 𝄞 —

The electro-chemical language of neurons (Figure 3.21) is the source of all communication and coordination within the body, giving rise to every action, thought, or unconscious process you have ever experienced. The expansive network of a total of one hundred billion (10^{11}) neurons, which speak this language, is responsible for perceiving, integrating, and transmitting information throughout the body. Understanding the process of communication in the nervous system requires an understanding of the structure and function of neurons.

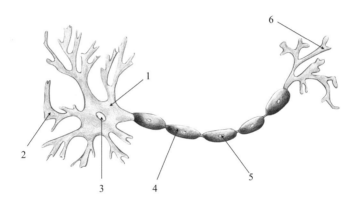

Fig. 3.21: Scheme of a human neuron: 1. dendrite; 2. cell body; 3. nucleus; 4. a part of the axon, with myelin; 5. Schwann cell; 6. axon terminal. Drawing by Yan Pang.

Most neurons are not like the highly specialized hair cells in the ear. Unlike the hair cells of the ear, neurons communicate through *action potentials*. Action potentials are an all-or-nothing electrical response elicited from a neuron. Let's examine how a neuron is able to create and communicate these action potentials.

In order to receive, integrate, and send a great variety of communication, neurons are equipped with highly specialized structures. First are the tree-like projections at the body of the cell, called *dendrites*. The dendrites receive information from other neurons and pass it onto the cell body. The cell body integrates this information and decides whether or not to send an action potential. Once an action potential is sent, it travels along the *axon* (a long, wire-like channel of the cell). Axons must carry the action potential to wherever the next neuron is, meaning that they can range from microscopic sizes to 2 feet long! The end of the axon is called the *axon terminal*, which contains pouches of chemicals called *neurotransmitters*.

Neurotransmitters are chemicals that influence the behaviors of neurons. There are over 60 different neurotransmitters, each with a slightly different function. We will discuss a few specific neurotransmitters (and their effects on emotion) in a later section (see Chapter 4.1.1.3). While action potentials make up the electrical language of neurons, neurotransmitters make the chemical one. When an action potential reaches the axon terminal, the cell releases specific neurotransmitters into a small space between itself and the next cell. This space is called the *synapse*, see Figure 3.22. Neurotransmitters cross the synapse and bind to receptors in the neighboring cell.

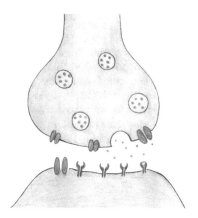

Fig. 3.22: Synapse. Drawing by Yan Pang.

These *receptors* on the post-synaptic cell allow it to receive neurotransmitters. Once a neurotransmitter is received by the post-synaptic cell, it activates a graded potential. *Graded potentials* are changes in the charge of a cell.

The voltage of a neuron is controlled by *ion channels*: proteins that allow certain ions into the cell while preventing other ions from entering. The resting voltage of a neuron is 55 millivolts (mV). An excitatory neurotransmitter brings the cell's charge closer to threshold, while an inhibitory neurotransmitter lowers the charge away from the threshold. Then, once the threshold is reached, the cell fires an action potential and the process continues on in other neurons until one neuron communicates to a muscle or gland (see Figure 3.23). After all of that, this neural information is finally expressed in some form of behavior at last!

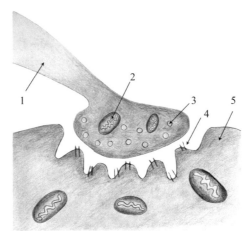

Fig. 3.23: Scheme of a human neuromuscolar junction: 1. end of the motor nerve with presynaptic terminal; 2. mitochondrion; 3. synaptic vesicles; 4. acetylcholine receptor; 5. muscle; muscular cell membrane: sarcolemma. Drawing by Yan Pang. We can see such connection also as a bridge between the reality of the mind, with thinking transformed into electric impulse of nerves, to the physical/physiological reality of movements in space and time of human body, moved by muscles. See the chapter about gestures and embodiment for these concepts.

3.2.3 Neuroplasticity

When neurons fire they communicate with each other. Extended communication between two neurons results in the development of more connections between the two, meaning that they increase their dendrite-to-synapse ratio. Neurons that do not fire with one another do not form new connections, and eventually disconnect from one another. These are two concepts of Donald Hebb's theory of synaptic neuroplasticity. *Neuroplasticity* is the brain's ability to change itself through the connection and disconnection of neurons. Hebbian theory has proven to be one of the most influential theories in all of modern neuroscience, guiding much of the field's research to this day [25]. Neuroplasticity is the basis of learning and memory: when we experience something we form new associations at the neural level. These new associations are facilitated by synaptic plasticity.

3.2.4 Music and the Brain Lobes

Despite the brain's ability to change itself, it seems as though certain functions of the brain are largely restricted to certain areas. Higher-order tasks of human cognition (such as music) have their functions mostly concentrated in the neo-cortex, although the emotional brain in the archicortex also plays a role related to emotions in music (see Section 4.3). There are four main lobes to the cortex

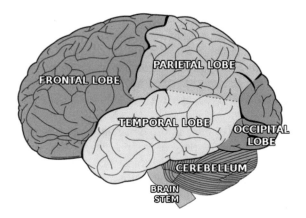

Fig. 3.24: The four neocortical lobes of the human brain, together with the brain stem and the cerebellum.

of the brain: the frontal, parietal, temporal, and occipital lobes (see Figure 3.24). Each of these lobes have some designated functions, although communication across the entire brain is necessary for most thoughts and actions. The *parietal lobe* has a bit of a hodge-podge collection of different functions. It contains the *somatosensory cortex*, which processes perceptual information regarding touch, pain, and temperature, but it is also tangentially involved with language processing. The *temporal lobe* is primarily involved in processing auditory information. It contains Wernicke's Area, which is responsible for language comprehension. The *occipital lobe* processes visual input and provides us with one, uniform understanding of the visual world around us. Finally, the *frontal lobe* is involved with high-order thinking, judgment, and self-control. Speech production is taken care of in Broca's Area in the posterior frontal lobe. The frontal lobe contains the *motor cortex*, a part of the brain that organizes and controls the parts of the body.

As we have previously discussed, music is first translated into electro-chemical information via the cochlea's hair cells. Once this information is received by the neurons, it is sent to the auditory cortex in the temporal lobe of the brain. The auditory cortex translates this information into what we perceive as pitch. After this point, it is unclear exactly how the brain is able to translate pitch, rhythm, articulation, and other aspects of music into a profound cognitive and emotional experience. However, it is clear that listening to music affects many parts of the brain, influencing parts of the brain responsible for pleasure (i.e. the nucleus accumbens) and emotion (e.g. the amygdala and hippocampus).

The amazing neurological influence of music is not restricted to music listening, but extends to music playing and learning. In fact, learning an instrument produces significant changes in the brain that can change one's behavior and ability. For example, Oechslin et al. [54] found that individuals who participated in music lessons for a number of years were found to have a greater hippocampal volume than those who did not take lessons. Oechslin and his colleagues also found that individuals with higher hippocampal volume (i.e. those who participated in music lessons) tested higher on creative aptitude tests. The researchers attributed increased creative aptitude to the increased hippocampal volume.

Music is not only a fascinating topic of neuroscientific study, but (due to its ability to activate a wide variety of brain areas) it also plays an important role in teasing apart different parts of neuroscience and making new discoveries. Music is not only a sensationally rich stimulus. It has an astonishing ability to access and trigger our thoughts and emotions. Given its omnipresence in the brain, surely we can say that music is an intrinsic component of the human experience.

4

Psychological Reality

Summary. The psychological reality of music is involved with the way that humans perceive and react to music. In this section we discuss the prominent theories regarding the relationship between music and emotions, the ways in which music interacts with psychopathology, and how human reactions to music can be measured.

$$-\,\oint\,-$$

4.1 Emotions and Music

Summary. Emotions and music seem to be inextricably intertwined. Especially since the advent of music in movies, music seems to confer emotions and emotions seem to confer music. When you listen to Barber's *Adagio for Strings* you may feel somber or sad. Alternatively, you may find yourself humming joyfully after having received a perfect score on an exam. But does music directly mirror emotion? And how do we classify emotions in a scientific way? In this section we will present and criticize the major theories about the relationship between emotions and music.

$$-\,\oint\,-$$

John Sloboda and Patrik Juslin provide an excellent explanation of the psychology of music [65]:

> Psychology is concerned with the explanation of human behavior:
> Behavior includes overt action as well as "inner" behavior, such as thought, emotion, and other reportable mental states. It can include behavior of which the agent is not fully or even partly aware, such as the dilation of the pupils of the eye.
> A psychological approach to music and emotion therefore seeks an explanation of how and why we experience emotional reactions to music, and how and why we experience music as expressive of emotion.

© Springer International Publishing AG 2016
G. Mazzola et al., *All About Music*, Computational Music Science,
DOI 10.1007/978-3-319-47334-5_4

In order to explore how music affects emotion, we first need to ask: "What is emotion?" Here we will present three basic approaches to define emotion.

4.1.1 Defining Emotions

There are a number of different ways to measure human behavior, spanning across a variety of different fields.

4.1.1.1 Lazarus and His School [35]: Categories of Emotion

Richard Lazarus focused much of his research in appraisal theory, which began with Magda Arnold. Appraisal theory emphasizes the importance of cognition as a precursor to emotion, that is, thought precedes physiological arousal that causes emotions to be felt. More specifically, appraisal theory suggests that the cognitions about an event are concurrent with the interpretation of that event. The way in which an individual interprets the event then gives rise to emotions. If they interpret the event in a positive light (that is, they made progress towards their "end-goal"), they would be more likely to be skewed towards positive emotions.

In the table below we define three different ways to measure emotional behavior.

Characteristics	Examples
Self-report	Feelings, Verbal Descriptions, Checklists, Rating Scales, etc.
Expressive Behavior	Facial Expressions, Gestures, Vocalization, Muscle Tension, etc.
Physiological Measures	Blood Pressure, Skin Conductance, ECG, EEG, etc.

Emotion is as much of a subjective experience as it is an objective one. That is, emotion is experienced differently by different individuals. As such, we cannot depend solely on objective measurements alone. In order to measure things that are unique to an individual, we use *self-report* methods. Self-report methods for emotions include verbal descriptions, rating scales (e.g. Likert scale), questionnaires, and inventories. Through these measures, we can collect data about individual experiences and organize them into broad categories. For example, if we wanted to look at whether Beethoven's *Fifth Symphony* could cause somebody to feel anxious we may have them listen to the music, then answer a Likert-scale question like the following:

Rate the extent to which you agree to the following:
I am feeling anxious. (Strongly Disagree) 1 2 3 4 5 6 7 (Strongly Agree)

Using the Likert scale we can assess the strength of an emotional value based on participant responses. While self-report measures allow us direct insight into a person's mind, they are not without drawbacks. Self-report measures are scientifically weak for two primary reasons.

First, in order to have high validity and reliability (that is, in order for them to consistently test what they are supposed) self-report measures must be long and repetitive. This is necessary because emotion is subjective and the interpretation of "I am feeling anxious" may vary greatly between one person and another.

Second, introspection is often not sufficient because there is a lot of thought that is unconscious in nature. For example, someone may not realize that they are nervous, or may underestimate their nervousness. People may misidentify, underestimate, or overestimate their emotions despite even visible responses to emotion that can clearly be seen from the observer's perspective.

In order to counter this second problem with self-report measures, we can observe *expressive behavior*. Expressive behavior can be described as the physical communication of thoughts, emotions, and feelings. This includes facial expressions, gestures, and vocalizations among many other conscious and nonconscious behaviors. Measuring expressive behavior is particularly useful for observing unconscious behavior, but it is also useful in confirming information from or pointing out inconsistencies in self-report measures.

Let us return to our question of whether Beethoven's *Fifth Symphony* can invoke anxious feelings in an individual. While playing the piece and observing the participant, we can use measures of expressive behavior (such as stressed faces or tensing of muscles) to determine if the participant is experiencing anxious feelings. While there is much support for measures of expressive behavior as highly valid and reliable measures of emotion (e.g. facial cataloguing and predictive value, investigated by Silvan S. Tomkins and Paul Ekman [66]), they too have two major flaws. Like self-report measures, expressive behaviors are still subjectively analyzed (granted, with less variation since it is usually the same observer over many participants). Additionally, while they are excellent at determining external behavior, they cannot account for internal behavior. Fortunately, the final of our three measures of emotions covers this aspect.

Physiological measures of emotion are perhaps the most convincing of measures, because they directly measure responses in our body. These responses are not biased by the participant's inability to accurately identify an emotion, or willingness to openly lie about it. Physiological measures are based on measurements of biological activity such as heart rate, cortisol levels, skin conductance, and brain activity and thus they are easily communicated through graphs and numbers. The greatest strength of physiological measures is that they are almost perfectly objective, so there is little to no possibility for errors in data collection on behalf of the participant or the experimenter. Unfortunately, physiological measures do have certain weaknesses. Physiological arousal for two different emotions can be identical, despite the fact that the emotions are vastly different. For example, stress hormones (such as cortisol) will be released into the blood stream when a person experiences anxious or nervous feelings. However, these hormones may also be released for other reasons, such as caffeine consumption. In short, while physiological measures tend to be highly accurate in regards to biological arousal, they do not always directly measure emotional arousal (or at least not the specific emotion that is being examined).

The ambiguous nature of physiological measurements makes it difficult to pinpoint the exact emotion a person is feeling, without feedback from the individual themselves or from observers (i.e., the experimenter). Additionally physiological measures can only be collected in certain laboratory settings and can be more time consuming or uncomfortable for the participant (making a study less attractive, and thus less likely to meet the preferred number of subjects).

Later in this chapter, we will discuss the use of *electroencephalographs* (EEGs) as a means of measuring emotion. EEGs measure brain activity by monitoring the difference in voltage caused by action potentials. This information is collected by attaching electrodes to the subject's head and the information is displayed on a computer monitor. There are many different neurons in the brain at once, but they fire at different frequencies. Additionally, these frequencies can be related to different functions and areas of the brain. Thus, by looking at the trends in neural firing at different bands of frequency, scientists can make assertions about the processes occurring in the brains of a participant at any given time during the EEG. We will discuss this in further depth in Section 4.3.

4.1.1.2 Russell and Barrett: Core Affect

Unsatisfied with the contemporary models of emotion that did not take human sociality into account, James A. Russell and Feldman Barrett (1999) [61] developed a model of emotion that acknowledged the role of activation in emotion formation and recognition. As part of their model, they referenced the neurophysiological concept of *core affect*. Psychologists used the term "affect" to describe emotions, feelings, and attitudinal valence (whether someone views something as good or bad). Core affect is a neurophysiological state measured on two continuous dimensions: pleasure versus displeasure and high activation versus low activation.

According to the definition provided by Russell and Barrett, core affect does not account for "prototypical emotions", or emotions that depend on direction towards an object. By this definition, core affect does not explain behaviors that require complete awareness of and planning for a situation, such as being in love with someone or being afraid of something. Instead, core affect models emotions that arise in a less conscientious setting such as the stress one feels at the end of a work day, or the elation one feels after finishing a project. The core affect theory of emotions suggests that the recognition and categorization of non-directed emotions arises from the two dimensions of pleasure and activation (see Figure 4.1).

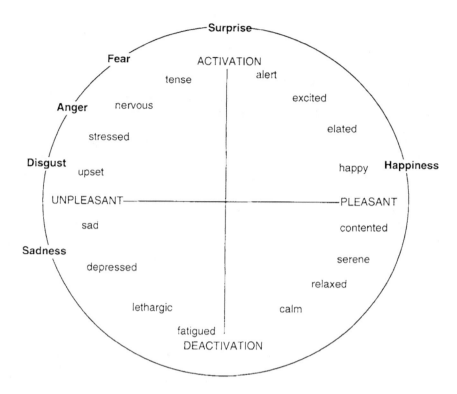

Fig. 4.1: Russell and Barrett suggested that emotions can be represented in a two-dimensional plane with two intersecting axes representing two dichotomies of emotion. On this plane, the horizontal axis represents the degree of pleasantness (pleasure/displeasure), while the vertical axis represents the level of arousal (activation/deactivation).

In many ways, the beauty of the Russell-Barrett model is its concise representation of how emotions can be formed and categorized. However, due to its simplicity, the Russell-Barrett model is not an adequate model for understanding the wide world of emotions. The authors themselves noted that "not all of the properties of our current category system *per se* are desirable from a scientific point of view. Vague boundaries, an unspecified number of overlapping categories, and ill-defined concepts limit the precision and rigor aimed at in science" (Russell and Barrett, 1999, p. 808).

Additionally, this model does not account for the experience of seemingly contradicting emotional states at once. For example, after brilliantly finishing a concerto concert, a violinist may be simultaneously happy and fatigued, yet the core affect model shows these emotions as being entirely separate. Russell and Barrett defend this by suggesting that these two emotions are being felt separately, but at the same time, and that core affect can be used to determine each of these emotions. However, the core affect theory itself does not explain how two emotions can arise at once. After all, if someone has no energy and feels slightly unpleasant (fatigue), how can they also have energy and feel very pleasant (happiness)?

Additionally, the core affect theory does little to explain the magnitude of emotions. That is, while it can categorize the emotion of stress as being moderately activated in an unpleasant way, it cannot explain the intensity of this emotion.

Music is very clearly capable of activating conflicting emotions. For example, someone may love listening to a particular song that reminds them of a past lover, inducing feelings of happiness, nostalgia, sadness, and longing all at once. Such a complex mixture of emotions cannot be explained by the core affect model. While it may be adequate for defining emotions arising from a stimulus that is consistent in its valence and magnitude, core affect is too simple and strict to explain complex emotions that are aroused by multiple stimuli, or by one stimulus with multiple interpretations. In order to understand the effects music has on emotions, a more flexible model should be used.

4.1.1.3 Mazzola: Neurotransmitter Model

There are nine primary types of neurotransmitters, each with a specific function/effect, see Figure 4.2. The concentration of different neurotransmitters gives rise to different emotions. When the concentration of one neurotransmitter is high relative to the presence of other neurotransmitters, the feeling (e.g. alertness) of that emotion will be more clearly felt, for there is little contamination by other emotions because of low concentrations.

However, when concentrations of different neurotransmitters are relatively similar, we may feel more complex feelings. For example, high levels of epinephrine paired with high levels of serotonin may result in euphoric energy, such as when you engage in an activity that you enjoy. In contrast, low levels of epinephrine paired with high levels of serotonin may result in a feeling of tranquility or contentedness, as experienced when one can finally relax and read a book at the end of a stressful day.

Thus a classification of emotions could be given by the activity of neurotransmitter languages.

4.1.2 Langer's and Gabrielsson's Thesis

Now that we have discussed how to define emotions, we can discuss their relationship with music. It is quite evident that music yields a special capability to activate our emotions. Whether it be the passionate anger incited by *Mars* from Gustav Holst's *Planets Suite*, the plaintive, sighing melody from Samuel Barber's *Adagio for Strings*, or even the tacky, yet motivational theme from *Rocky* called *Gonna Fly Now*, we can all agree: music affects us in amazing ways.

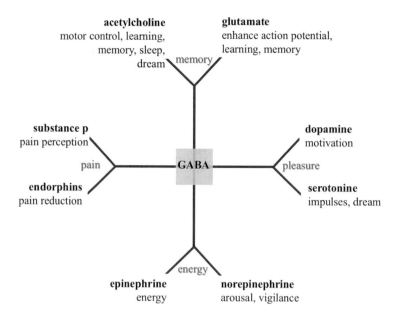

Fig. 4.2: Mazzola's model of emotions, based on the functions of the neurotransmitters. Note that different fields of emotion (i.e., memory, pain, pleasure, energy) can interact with one another based on levels of concentration.

Of course, given the magnitude and variety of its effects, understanding the interaction of musical structure and human emotions has become a subject of intense scrutiny in the scientific and philosophical communities. One of the most important topics in this area of study, is the degree to which emotions and musical structure resemble each other.

Amongst the muddle of opinions generated on this topic, the controversial and polarizing theory of isomorphism, proposed by Alf Gabrielsson and Susan Langer [20], has emerged as a powerful force in shaping future theories. *Isomorphism* refers to the idea that two structures covary such that variation in one causes a corresponding functional change in the other. As Langer puts it, "music is a tonal analogue to emotional life."

Gabrielsson took Langer's idea and expanded on it, claiming that "emotion, motion, and music are isomorphic" [20]. He extends into specific examples of isomorphism in music. For example, Gabrielsson suggested that notes with a legato (i.e., long and drawn out) articulation notes create a feeling of solemnity, while notes that are light and jump around tonally inspire happiness.

However, the authors of this book are highly skeptical about the validity of this idea. For one, with 72 notes in Western music, there are

$$2.23 * 10^{36}$$

combinations. That means 2,230,000,000,000,000,000,000,000,000,000,000 possible combinations! With so many possible combinations of notes, it is unlikely that they form exact correlations with emotions. However, given that emotions can be graded on a continuous scale, this may be possible, but a second problem is approached. While Gabrielsson's thesis may work generally (e.g. fast pieces tend to be happier) they do not work specifically (i.e., not all fast pieces are happy, many are agitated). Additionally, the same piece of music can cause different people to experience different emotions. If music were isomorphic with emotions, how could people have vastly differing interpretations of the same piece? See also our theory on this effect at the end of Section 4.3.

4.2 Measuring Electrical Responses to Music

Summary. Electroencephalographs (EEGs) measure the rate of production of action potentials in the brain. They are not specific enough to detect when a particular neuron fires, instead they measure voltage fluctuation over the entire brain. However, since neurons in different parts of the brain fire action potentials at different frequencies and at different times, information collected by the EEG can be separated into six different frequency bands corresponding to activation in different areas of the brain. This, of course, results in a broad data set that is not very representative of a specific function in the brain. In order to make assertions about which parts of the brain are being activated, one can calculate the average activity and figure out which band is most prominent. The resulting product is called an *event related potential* (ERP).

4.3 Some Physiological Evidences

We should now give a number of empirical evidences of the neurophysiological emotion-related responses to music, so we are relying on the third characteristic, physiological measures, in the above definition of emotions. A number of neurophysiological experiments have been done where the electrical activity of the human brain was investigated in its response to acoustical inputs of musical structures. The first two of them revealed significant differences between female and male listeners. In 2003, Stefan Koelsch et al. [30] measured event-related brain potentials (ERP, short event-induced spikes delayed by 300 to 600 msec) taken from electrodes on the surface of the skull (Figure 4.3). The chosen population were 5- to 9-year-old girls and boys without musical training. The musical stimuli were three short cadential sequences of chords, each in two variants. The first was the typical $I - IV - V - I$ cadence, with variant $I - IV - I - V - I$. The second was altered to take the Neapolitan chord instead of the last I, or the middle one in the variant, and the third showed a cluster instead of the final I or the middle I in the variant. The distribution of the ERP over the skull is shown to the right in Figure 4.3. We see that the girls have a symmetrical activity map, whereas the boys show a maximum of activity in the right frontal brain[1].

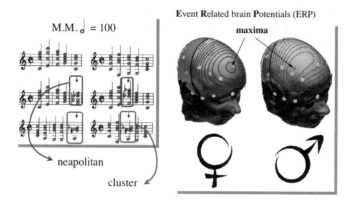

Fig. 4.3: Stefan Koelsch's experiment with event-related potentials as a response to three cadential chord sequences.

In his extensive studies [57] of EEG responses to musical stimuli from 1986 to 1998, Hellmuth Petsche et al. obtained quite detailed information about the localization and coherence of surface EEG activity in the different classical frequency bands, namely $\theta = 4 - 7.5$ Hz, $\alpha = 8 - 12.5$ Hz, $\beta_1 = 13 - 18$ Hz, $\beta_2 = 18.5 - 24$ Hz, $\beta_3 = 24.5 - 31.5$ Hz, see also Figure 4.4. His brain maps show power ("Leistung"), local coherence ("lokale Kohärenz"), and interhemispheric coherence ("interhem. Kohärenz"). Coherence means that they measured how strongly electrical activities in different localities of the brain are coupled to each other, and this is a sign for simultaneous, and therefore coordinated, processing of musical stimuli in different brain areas.

Petsche presented the first movement of Mozart's string quartet KV 458 to 24 male music students and to 28 female music students. So the response has to be interpreted as a global reflection of the musical structure. The results (Figure 4.4) show that interhemispheric coherence, in particular in the beta band, which is correlated to intellectual brain tasks, is much stronger (red color) for females than for males. This confirms the well-known fact that females have a stronger general interhemispheric activity than males. We also see the significant local

---

[1] We are aware that the big and important topic of gender in musical composition, performance, and theory is not dealt with in this book. This has two reasons: To begin with, the field would deserve a very cautious discourse on the different rationales for gender differences in music. Since this book is an introduction, we do not delve into this complex field.

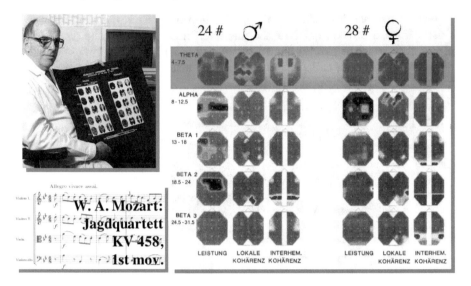

Fig. 4.4: Hellmuth Petsche's experiments with EEG responses in different frequency bands, for male and female populations, to the first movement of Mozart's string quartet KV 458.

coherence in the right hemisphere on the β_2 band, which confirms that the right hemisphere is an area of holistic music processing, typically needed for the recognition and evaluation of complex shapes, like motives and melodies. The θ band is neglected because it can be affected by signal noise.

The findings of Petsche are quite sensational in view of the question about the nature of cognitive music processing in the right cortical hemisphere. Since Petsche's findings confirm cognition and evaluation of complex shapes via local coherence in the right hemisphere, we might ask what happens if this coherence is impaired by some physiological conditions. One of those impariments is reported from anatomical measurements of brain weights in schizophrenics. It has been shown that these patients suffer from a significant loss of axon mass in the right hemisphere [38, Chapter 1], axons being the connectors between neurons, see Figure 3.21. This implies that the local coherence in the right hemisphere is lower than for normal humans. This has been confirmed by EEG investigations similar to Petsche's coherence measurements. The labyrinthic existentiality of schizophrenics appears as a consequence of the thinned-out network of neurons: They have to run through a labyrinth of axons instead of performing on a dense neuronal maze. The linearization of reality is evident in the art of schizophrenic persons, such as the graphical music work of Swiss schizophrenic Adolf Wölfli (Figure 4.5).

All of this should also have consequences for the relation of schizophrenic persons to music. It would imply that they have a lowered performance in creating, recognizing, and processing complex musical shapes. In particular, this

Fig. 4.5: Swiss schizophrenic artist Adolf Wölfli produced a large number of drawings of musical topics, mainly score-like displays. The photography shows him with one of his paper trumpets.

would suggest that there are no great schizophrenic composers, who are required to imagine complex shapes while creating high-ranked works. It is in fact true that no such composers are known. The only critical case was Robert Schumann, but it has been proved that he was not schizophrenic but suffered from a tumor on the skull base [64]. Other psychoses, such as depression, do not affect musical creativity, as is, for instance, beautifully and tragically shown by the example of Peter Tchaikovsky. Of course, it is by no means a negative judgment about schizophrenics to put into question their compositional abilities. It might also be true that in future times they could compose a type of music that has high qualities that we cannot view and appreciate at present.

A third example of neurological evidence of musical stimuli, and this time strongly related to emotion, is Mazzola's joint research with epileptologist Heinz-Gregor Wieser at the University Hospital in Zürich from 1984 to 1986 [37, 68, 69]. They were in the interesting position to have depth electrodes implanted in humans with chemically intractable focal epilepsy for presurgical evaluation (localization of the focus). During this evaluation, it was possible to present musical structures to the patients via headphones and to have the depth structures of the brain respond to this input. For medical reasons, a number of electrodes were positioned in the hippocampal formation of the limbic system, which is a prominent structure of the emotional brain (situated below the neocortical brain layers). Figure 4.6 shows the electrodes in X-ray imaging and the EEG derived from these electrodes in the range of some 50 μV. We see the input score, which is an example of first species Fuxian counterpoint,

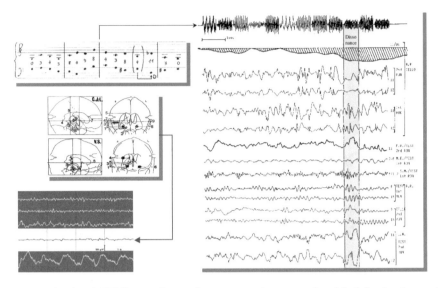

Fig. 4.6: The depth EEG recordings of a counterpoint example with defective interval (dissonant major seventh replacing the major third consonance). The depth EEG from the hippocampal formation of the emotional brain shows a significant disruption when the dissonance appears.

on which we have changed one of the consonances (a major third) to a major seventh, a strong dissonance in the third but last position. The sequence of a number of EEG waves reacting to this dissonant disruption in the consonant context is shown to the right. We see a significant change in the EEG structure when the dissonance appears. This proves that beyond the conscious perception of dissonances versus consonances, this EEG disruption is produced in the deep structures of the emotional brain. The patients in these tests were normal European male adults with common musical taste and no instrumental education. *The result of this investigation is that in the deepest structures of the emotional brain (in the archicortex positioned below the neocortex), basic musical structures of consonances and dissonances are significantly distinguished from each other on the level of EEG waves.* We therefore have strong arguments for the emotional effect of music beyond the conscious self-report or behavioral response.

It is remarkable that the hippocampal formation is understood as a gate to the subconscious. Therefore music may act as a key (individual for each person) to subconscious contents, also referred to in Claude Debussy's prélude *La Cathédrale engloutie* (The Submerged Cathedral).

4.4 Psychopathology and Music: van Gogh, Wölfli, Harrell, Tchaikovsky

Creative minds tend to view the world in new and unconventional ways, and thus often seek out intense and unstable circumstances, even if those experiences are simply within their own mind. This leads to a significant connection between art, the act of creation, and psychopathology itself.

Some of the most creative ideas are generated during chaotic mental states characterized by loosening of normal associations, which resemble the psychosis of mania or schizophrenia. The act of creation can be intertwined with these mental states. Perhaps the best example of such a connection is Robert Schumann, a 19th century German composer, widely regarded as one of the best of his time. Suffering from

Fig. 4.7: Tchaikovsky (1840-1893).

manic depression, Schumann would go through multiple swings throughout his years, from states of hypomania, under which he would be extraordinarily prolific, to states of depression, where he was unable to produce any compositions at all. For Schumann, and many others, this was motivation enough to not seek out treatment, for fear of losing the manic states, which allowed for him to produce such complex and abundant melodies. These states of mania are conducive to creating some of the greatest artistic endeavors to this day.

For those suffering with mental illness, art also provides a release, and a way to communicate with the world outside of their own head. The unique capabilities and effects that exposure to the arts can have on an individual are undeniable. Artistic expression has been known to relieve stress, lift spirits, reduce pain, and subside even the most aggressive symptoms of mental illness. Music provides a temporary inner harmony and a way to ward off even the worst manifestation of mental illness. For artists like Vincent van Gogh (Figure 4.8) and Peter Tchaikovsky (Figure 4.7), who both struggled with persistent dysthymia, suicidal thoughts, and general despondency, work was their only means of escape from an other-

Fig. 4.8: Vincent van Gogh (1853-1890)

wise meaningless existence. It was their means of self-medication beyond alcoholism and appalling 19th Century medicinal techniques, such as blood-letting. Tchaikovsky said:

"I need work like I need air to breathe. As soon as I am idle, despondency overcomes me. I am dissatisfied with myself and even hate myself. Only work saves me. Without music I would go insane."

Tchaikovsky was highly depressive, also because he was homosexual and could not risk a coming-out in those days. He committed suicide after a dinner, drinking the cholera infected Newa river water.

This evidences the profound effect and necessity that many of those suffering from psychopathologies feel as a result of working with the arts. Their creation is absolutely essential to their sense of being, and is a certainty in a world full of ambiguity.

Fig. 4.9: Tom Harrell (1946-)

In contrast to depression, schizophrenia has strong influence on musical abilities. No schizophrenic great composer is known. We have a theory that connects schizophrenia with the abscence of higher musical abilities as a composer. Musical imagination is strongly tied to the spatiotemporal imagination that is located in the right hemisphere of the brain. Schizophrenia physiologically manifests in a measurable lower neuronal interconnectivity in the right hemisphere. This means that schizophrenics have great difficulties (longer neuronal paths!) building complex spatiotemporal shapes, as they are essential in musical imagination.

This lack of neuronal interconnectivity can also be understood as a kind of labyrithic structure, a system of lines instead of a "surface of connections". Vincent van Gogh, who suffered from schizophrenia, typically drew surfaces as systems of lines. Or Adolf Wölfli drew a great number of pictures full of musical scores that were embedded in labyrinthic line systems. One could therefore call schizophrenia the "minothaurus complex", the life in a labyrinth, in fact a classical existential mode for schizophrenic persons.

However, as well as providing a sense of meaning and release through the physical act and purpose of creation, music itself can also be a treatment. For Tom Harrell (Figure 4.9), a multi-award winning jazz trumpeter, the use of the creative process eliminates his symptoms completely. Having suffered from debilitating paranoid schizophrenia, Harrell hardly has known a day without the feeling that he is under attack, and a constant chorus of voices in his head. When he is making music, however, these all disappear, he can escape his labyrinth. The significant shift that Harrell experiences when creating music is a testament to the power and connectivity between his psychopathology and the music itself.

4.5 Renate Wieland on Gestures and Emotions

Renate Wieland hypothesizes that emotions arise from a gestural basis. She suggests that emotions begin as physical actions towards an object, but the physicality is lost in the process of internalization. Despite the intangible nature of emotions, Wieland argues that that emotions are inseparable from the physical actions and situations that they arise from and are expressed in.

This hypothesis is supported by psychological research in the importance of the face in emotion formation. This concept, called *facial feedback theory*, was perhaps first addressed by Charles Darwin [13]. In his work, Darwin suggested that physiological factors may influence emotions rather than act simply as a result of them.

Much later, in 1962, the social psychological researcher Silvan Tomkins provided the first in-depth exploration of this relationship [66]. While researching the muscles involved in different facial expressions, Tomkins and his students realized that they began to feel strong emotions corresponding to facial expressions they made. For example, after frowning for an extended period of time, the researchers noticed that they felt angry, despite the absence of any external stimuli. This lead Tomkins to explore the idea of facial feedback, suggesting that sometimes we infer our emotions from our facial expressions rather than vice versa. The proposed direction of causality, going from gestures (i.e., the facial expressions) to emotions is highly reminiscent of Wieland's theories.

Beyond the effects that gestures have on emotion, Wieland suggests that the very structure of the human consciousness is based on a gestural coordinate system. This system is both external (i.e., the ability to track and plot the movement of the body) and internal. While Wieland does not propose a mechanism for the internal coordinate system, she does provide a description of how it may work. She argues that these gestural coordinates determine the way in which thoughts unfurl and shape the gestural pathways in the physical world. In turn, these gestures activate similar internal coordinates in other people, resulting in some form of shared experience.

As an example of this theory in action Wieland points to the movements of a conductor.

"Erfahrbar werden musikalische Gesten in der freien dirigentischen Bewegung in der Spielbewegung und sublimiert in der geistigen Mimesis reiner Imagination. Gleich in welcher Ebene, immer sind es Experimente im Raum."

"Musical gestures of the conductor are experienced from the free directing movement in the playing movement and it is sublimated in the spiritual mimesis of pure imagination. Independently of the level, it is always about experiments in space."

Both Wieland's and Tomkin's theories are highly debated in psychology. Many researchers reject the direction of causality proposed by Tomkins, pointing out that his studies may simply demonstrate the previously known causational link between emotions and facial expressions. The exact underlying causes of emotions are still largely unknown, and will likely remain a mystery for psychologists, neuroscientists, and other behavioral researchers to examine for many years to come.

5

Mental Reality

Summary. We have seen how music lives in the physical reality and in the mechanisms of human perception and emotion. In this chapter, we will discuss how music lives in the symbolic dimension of score notation, the tuning spaces, and their mathematical symmetries for harmony and counterpoint.

5.1 The Role of the Mental Reality

Although music was understood as a symbolic field since the Pythagorean tradition in ancient Greece, the French philosopher René Descartes (Figure 5.1) claimed that music needs a psychological foundation. In his *compendium musicae* (1618) [14], he explains eight fundamental rules for making and understanding music, such as "music must be simple to please the soul."

Fig. 5.1: René Descartes (1596-1650).

In recent times, the mental reality of music has been rediscovered, and has been supported by the work of mathematician Leonhard Euler. In this chapter, we will discuss the mental reality of music that complements (and sometimes contradicts) physical and psychological aspects.

5.1.1 The Musical Score

The score represents a mental reality, neither physical nor psychological. Figure 5.2 shows notes and their correspondence to piano keys. Neither physical

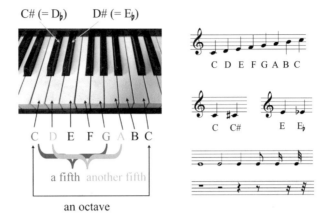

Fig. 5.2: Musical score notation (with note durations and corresponding rests) and its reference to a keyboard.

duration nor pitch are specified in the note symbol. This is dramatically shown in John Cage's composition *As Slow As Possible* (ASLSP), Figure 5.3. Cage's score is being realized in Halberstadt's church, where the piece is playing as we write this text (and probably as you read it). The piece began on September 5, 2001, and will continue to be played for 639 years. The beginning of the score is shown in Fig. 5.4. The first quarter notes lasted seventeen months.

Fig. 5.3: An ASLSP organ.

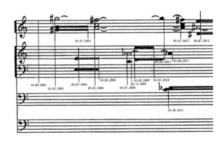

Fig. 5.4: Cage's ASLSP score.

5.1.2 The Pitch Space

Pitch is also not specified precisely in the score. It depends, on which instrument the score is played. In European music the selected pitches are described in a pitch space of frequencies, see Fig. 5.5.

The selected frequencies are multiples of a fundamental frequency, for example middle C, and powers o, q, t of 2, 3 and 5 that are rational numbers

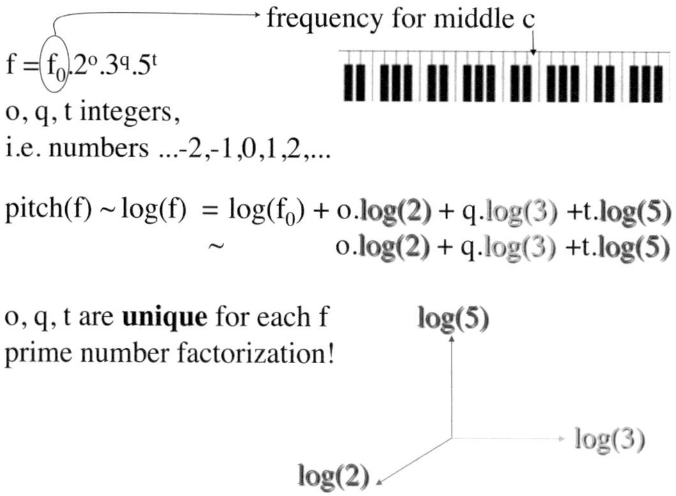

$$f = \left(f_0\right) 2^o . 3^q . 5^t$$

frequency for middle c

o, q, t integers,

i.e. numbers ...-2,-1,0,1,2,...

pitch(f) ~ log(f) = log(f$_0$) + o.**log(2)** + q.**log(3)** + t.**log(5)**

~ o.**log(2)** + q.**log(3)** + t.**log(5)**

o, q, t are **unique** for each f

prime number factorization!

log(5)

log(3)

log(2)

Fig. 5.5: Pitch space.

(such as $3/4$, $-2/3$ etc.):

$$f = f_0 . 2^o . 3^q . 5^t,$$

where the number 2 relates to the octave $2 : 1$, the number 3 to the fifth $3 : 2$, and the number 5 corresponds to the major third $5 : 4$. See Figure 5.12 for a list of all intervals within an octave. The three directions musically correspond to fundamental musical interval 'directions' as recognized by Gioseffo Zarlino, see[1] Figure 5.6.

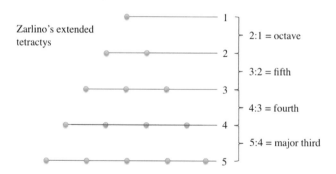

Zarlino's extended
tetractys

1

2:1 = octave

2

3:2 = fifth

3

4:3 = fourth

4

5:4 = major third

5

Fig. 5.6: Zarlino tetractys.

As the pitch is essentially the logarithm of frequencies, it is a linear combination of the logarithm of 2, 3 and 5:

---

[1] The Pythagorean tetractys only includes the first four rows.

$$pitch(f) \sim o.\log(2) + q.\log(3) + t.\log(5).$$

It can be shown that these three logarithms behave like three vectors in three-space, looking in different directions. This means that a pitch can be represented as a point in a three-space, see Fig. 5.5 for a graphic visualization.

5.1.3 Euler Space

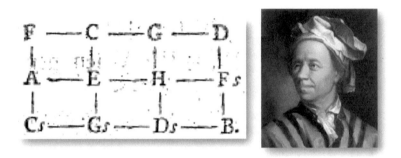

Fig. 5.7: Speculum and Euler.

In 1773, mathematician Leonhard Euler published a book, *Speculum musicae* [15], where he first used this spatial representation (Fig. 5.7), with fifths and thirds, but without showing the octave direction.

One octave of all twelve chromatic pitches in Euler space is shown in Fig. 5.8. In this space, every note is shown as a point with corresponding coordinates in octave, fifth, and third direction. For example, starting from pitch C, pitch G has $3/2$ of C's frequency. This means, when stepping over to logarithms, that we have to go one unit in fifth direction $\log(3)$, and minus one unit in octave direction $\log(2)$.

It is astonishing that this display looks quite irregular. Why would one choose such a configuration for the pitches of traditional European music? We shall see below that this irregularity is only apparent, there is in fact a very important hidden symmetry in this octave.

If we move to pitch classes, we see the projections of pitches into the plane of fifths and thirds, as shown in Figure 5.9.

5.1.4 Zarlino's Symmetry

In this representation we can see the relation between major and minor triadic chords following Zarlino's theory. Figure 5.10 shows that the minor chord is the symmetric image of the major chord using a rotation by 180 degrees.

In Fig. 5.11 the twelve chromatic notes of 12-tempered scale in Euler space are shown. They are much simpler in their arrangement as they lie on one line of the octave direction.

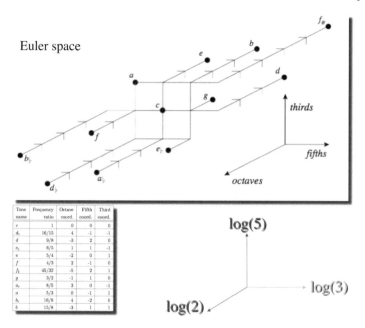

Tone name	Frequency ratio	Octave coord.	Fifth coord.	Third coord.
c	1	0	0	0
d♭	16/15	4	-1	-1
d	9/8	-3	2	0
e♭	6/5	1	1	-1
e	5/4	-2	0	1
f	4/3	2	-1	0
f♯	45/32	-5	2	1
g	3/2	-1	1	0
a♭	8/5	3	0	-1
a	5/3	0	-1	1
b♭	16/9	4	-2	0
b	15/8	-3	1	1

Fig. 5.8: Euler space.

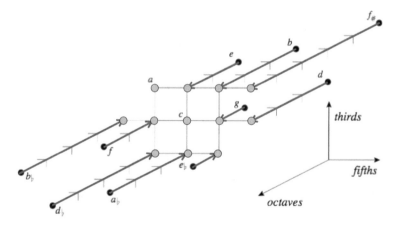

Fig. 5.9: Pitch classes.

5.1.5 The Hidden Symmetry of Counterpoint

Although the chromatic octave shown in Fig. 5.8 looks very irregular, it has a unique hidden symmetry, as shown in Fig. 5.13. This symmetry is important to understand musical counterpoint. What is counterpoint? In short, this is the art of composing music with a number of different voices. When one considers the intervals between (simultaneous) notes of different voices, coun-

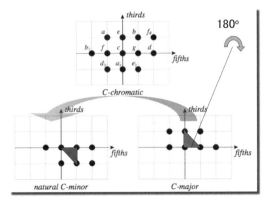

Gioseffo Zarlino (1517 - 1590): major and minor

Fig. 5.10: Zarlino's relation between major and minor.

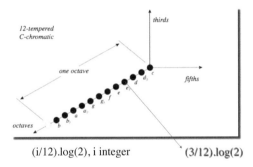

Fig. 5.11: Graphic description of the 12-tempered tuning.

terpoint essentially only allows consonant intervals, which are 0, 3, 4, 7, 8, 9 semitones (unison, minor and major third, fifth, minor and major sixth). See also our discussion of EEG and the perception of intervals in Section 4.3. The rules of counterpoint then describe which intervals can succeed which intervals. The forbidden dissonant intervals are: 1, 2, 5, 6, 10, 11 (minor and major second, fourth, tritone, minor and major seventh). Figure 5.12 show the list of all intervals from pitch C.

To understand this symmetry, which exchanges D and E, B and Db etc., as seen from the red arrows in Figure 5.13, we look at the circle of pitch classes, which are displayed similarly to a clock with twelve hours, i.e., the class of C goes to 0, the class of Db goes to 1, etc., stepping up by half tone steps. Looking at intervals from the musically important "leading tone" B in C-major tonality[2], this symmetry is realized by the symmetry formula $d = 5 \times c + 2$. This symmetry exchanges the consonant intervals (0, 3, 4, 7, 8, 9) and dissonant intervals (1, 2, 5, 6, 10 11), see Figure 5.14. For example, pitch D corresponds

[2] See our discussion of the concept of tonality in Chapter 8.

Fig. 5.12: All intervals from pitch C with the numbers of semitones.

to number distance $c = 3$ from B, the formula maps it to $5 \times 3 + 2 = 17$, which corresponds to 5 on the circle. Therefore consonance 3 (minor third) is mapped to dissonance 5, the fourth. You see that every consonance is mapped to a dissonance, and vice versa, by the given formula $d = 5 \times c + 2$.

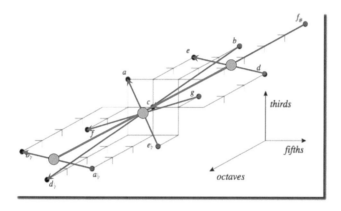

Fig. 5.13: The symmetry of the chromatic scale.

This formula is central to understand the rules of counterpoint [2], and we see that the chromatic scale has this symmetry in its hidden regular structure.

5.1.6 Euler's Gradus Suavitatis Function

Euler also tried to define consonances and dissonances by mathematical formulas, his *Gradus Suavitatis*, as shown in Fig. 5.15. This formula gives quite good values for the intervals in the chromatic scale. In his theory, Euler argued that the human brain would interpret (and substitute) any interval—being a deformed version of one of the just intervals—to the intervals having simple frequency ratios such as $2/1, 3/2, 4/3, 5/4$. However, as is shown in the Fig. 5.15, counterpoint has a different definition of consonant intervals (marked by blue triangles). This means that counterpoint theory does not follow the mathematical model used by Euler. So this is a clear example of the difference between symbolic and physical models of music. Euler's formula reflects the physical

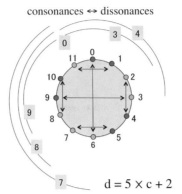

Fig. 5.14: Polarity between consonant and dissonant intervals. Identifying 0 with C, 1 with C♯, etc., we count intervals from 11 (B). The polarity maps consonant intervals (c) to dissonant ones (d).

interval relations (via those exponents that stem from physical considerations), whereas counterpoint diverges from this approach and has its own thoroughly symbolic perspective.

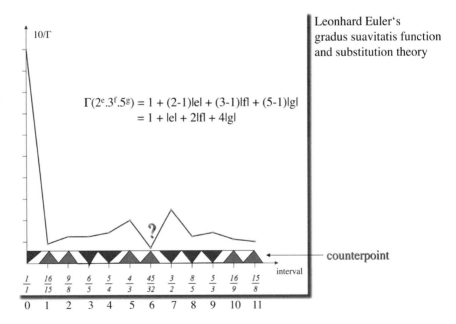

Fig. 5.15: Euler's gradus suavitatis.

Part III

Semiotics

6

Generalities about Signs, Neumes, Periods and Development Sentences

Summary. Semiotics is the dimension of meaning. It studies the structure of symbols and signs and their associated meanings. We will start by introducing the basic principles of semiotics: the different levels of symbols and the structure of musical symbolism. We will then discuss philosophers and linguists that studied semiotics, and explain their theories as they relate to music. We will apply the theories of Ferdinand de Saussure, who is perhaps the most influential scholar of semiotics, to analyze and exemplify music as a symbolic system. In an overview of more contemporary analyses of symbolism in music, we will discuss the HarmoRubette software and Riemann harmony. Additionally, we will discuss the Babushka Principle of semiotics, which leads into such topics as connotation, motivation, and metatheory.

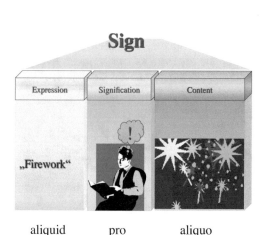

Fig. 6.1: The sign's anatomy.

6.1 Definition of Signs

Signs are sets of symbols with a conventional meaning that is associated to. For example, a musical score is a collection of symbols that are interpreted and translated into musical gestures and sound by professionally trained performers. A sign can be divided into three components: expression, signification and content. For example, when we read the word "firework", we read an expression, "firework", a set of characters, to which we associate a content (the sounding firework) via the signification process, see Figure 6.1.

6.2 Neumes

Neumes constitute an example of musical notation in Western culture that is historically derived from gestures of the choir conductor. They were developed in the context of Gregorian chant, the first Christian liturgical music. Gregorian chant was developed early in the Middle Ages and influenced by the Greek modes. It is a form of initially unaccompanied monophonic chant. The Western musical notational system is derived from Gregorian chant, whose origins, as we will see, stem from gestures. In 4th- and 5th-century Europe, there was no musical notation for melodies. Cantors in monasteries learned them by heart. Later, some signs for melodic shapes were added to words written in manuscripts, called *neumes* (from the Greek word for sign), see Figure 15.1. Initially, these neumes were positioned above syllables of the text, without any reference to a precise pitch. Successively, neumes have been introduced in a four-line musical stave, the ancestor of our modern five-line stave. In the context of semiotics, neumes are early musical signs that signify gestures. See our discussion of gestures in Chapter 15.

Fig. 6.2: The evolution of neumes to modern notation.

6.3 Musical Signs as Language

Analogously as it happens for an English written text, we can also subdivide a musical score into sentences, phrases, and periods. The musical equivalent of commas, colons and dots—periods are given by specific sequences of chords called *cadences*. When we look to a musical score, see Figure 6.3, we are observing its surface and its *expression*. When we start an analysis of it and we divide the score into periods and sentences and so on by making distinction between chords and sequences of chords, we are dealing with the *signification* of the score.

Finally, when we just look to the obtained information about the harmony of the piece (harmony is the part of music theory that studies the meaning of chords), we are dealing with its harmonic *content*. This is an example of how to view a score as a *sign* constituted by three parts: expression, signification and content.

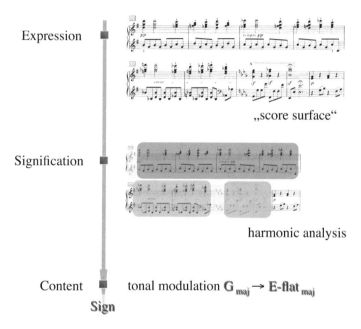

Fig. 6.3: A score's harmonic signification.

De Saussure and Peirce: the Semiotic Architecture

Summary. Semiotics has been studied by linguists and philosophers for many years. The first attempts to define the components of a sign system were made in 1865 by the United States philosopher, Charles Sanders Peirce. After Peirce came Ferdinand de Saussure, Louis Hjelmslev, and Roland Barthes, each with differing views on the components of a sign system. In this chapter we will explore contributions made by the four aforementioned semiotic theorists and discuss the semiotic architecture that their theories illustrate.

7.1 Pierce

The American scholar Charles Sanders Peirce (see Figure 7.1) was one of the first theorists to explore the world of semiotics. Though he studied chemistry during his college education, Peirce was truly a Renaissance man, publishing research in semiotics, linguistics, psychology, philosophy, statistics, and economics among other fields. Peirce's philosophical analyses spanned from concepts of logic to theories about the mind. Semiotics was born from Peirce's desire for a system of analysis to describe the general use of signs, as inspired by their application in logic.

However, it is important to note that Peirce's contributions to the field of semiotics were in the form of what he called "semeiotic", which is fundamentally different than the field of semiotics that we know today. Peirce's studies focused on the existence of the "triad of representing relation," in which a symbol was comprised of an *object* (something being represented), a *representamen* (something different than the object that represents the object and is related to a third concept), and the *interpretant* (the concept that the representamen appeals to). However, this proposal was complex, lacking clarity as to where the representation began and the represented concept ended. Later semiotic

© Springer International Publishing AG 2016

G. Mazzola et al., *All About Music*, Computational Music Science,

DOI 10.1007/978-3-319-47334-5_7

scholars sought a more eloquent, if not simpler, model for symbolic representation.

Fig. 7.1: Ch. S. Peirce (1839-1914).

Fig. 7.2: F. de Saussure (1857-1913).

Fig. 7.3: L. Hjelmslev (1899-1965).

7.2 de Saussure

While Peirce laid the foundations for the field of semiotics, Ferdinand de Saussure is the true founding father of modern semiotics. Ferdinand de Saussure was a Swiss linguist and philosopher (see Figure 7.2). Saussure established himself as a brilliant linguist, when he published his first and only book *Memoir on the Original System of Vowels in the Indo-European Languages* as a student. Saussure was not a particularly prolific writer, but his few publications revolutionized the field of semiotics. Moreover, Saussure as instructor and professor had an immense influence in the field. His most influential work was a culmination of his lectures grafted together by his students and colleagues, published post-mortem in 1916. In this short introduction to linguistics entitled *Cours de Linguistique Générale*, Saussure proposed six dichotomies (to be discussed in Chapter 9) that describe the nature of linguistics or, more generally, semiotics.

In a structure reminiscent of Peirce's argument, Saussure suggested that symbols are comprised of three interacting components: the *signifier* (the symbol used to symbolize an object), the *signification* (the process of symbolization), and the *signified* (the object being symbolized). Though it has been modified by linguists such as Louis Hjelmslev and Roland Barthes, this theory forms the backbone of modern semiotics.

7.3 Hjelmslev

Louis Hjelmslev was a Danish linguist who, like Saussure and Peirce, was interested in semiotics (see Figure 7.3). Hjelmslev is best known for his theory of glossematics, which revised Saussure's version of semiotics and added new concepts. Of these new concepts, the most important was what is called "connotation", which will be described in Chapter 9.3. In addition to this, Hjelmslev revised the signifier/signification/signified structure that Saussure proposed. In his book *La stratification du langage*, Hjemslev instead proposed the structure of *expression* (the symbol), *relation* (the relationship between the symbol and the content, similar to the signification process), and *content* (the meaning expressed by the symbol).

While Saussure built the backbone of semiotics, Hjelmslev was responsible for much of its development and popularization. Hjelmslev's theories allowed semiotics to expand vastly, branching out to become useful in areas outside of linguistics and philosophy within which they had largely been contained. In fact, Hjelmslev's contribution of connotation explains the infinite expansion of sign systems that Peirce originally desired to capture with his object/representamen/interpretant model of semiotics.

For the purposes of this book, we will use a model of the semiotic architecture that combines the ideas of Hjelmslev and Saussure. The combined sign model is presented as expression, signification, and content. This model incorporates the ideas of expression and content from Hjelmslev's model, while maintaining signification from Saussure's model. We feel that this is the clearest and most intuitive explanation of the semiotic architecture for the English language. Given this model, we analyze music (and eventually expand out to more general areas), a vastly different topic than the traditional use of semiotics as a linguistic tool. Today, it is common to use semiotics to study sign systems in a wide variety of areas, but this was not always the case.

7.4 Barthes

Roland Barthes was one of the first scholars to expand the use of semiotics beyond the linguistic realm. As a cultural analyst, Barthes used semiotics to explain the meanings of behaviors and cultural symbols. Barthes didn't stop there, though. He extended the use of semiotics to gastronomy, fashion, sociology, and even traffic signs. He was truly the first individual to act on the idea that semiotics can be used to study everything. Barthes was largely responsible for the structuralist movement of semiotics (a movement based on Saussure's ideas). In his book *Elements of Semiology* [3], Barthes suggested an anatomy of semiotics (see Figure 7.4). This anatomy of semiotics allows us to show how semiotic elements interact across different axes. Syntactics, for example, are located on the horizontal axis and are defined as the juxtaposition of different signs (as in a sentence or musical phrase). Semantics, the process of choosing how to get from expression to meaning, is represented on the vertical axis. Finally, pragmatics, or the usage of signs, is represented by the signification component of the semiotic architecture. In this way, Barthes was able to

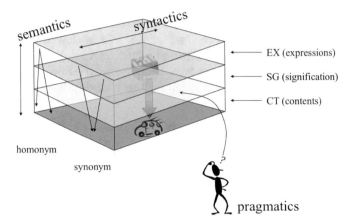

Fig. 7.4: The Semiotic Anatomy as suggested by Roland Barthes (1964). Note the syntax as the horizontal axis and semantics as the vertical axis. Note too: a synonym is a word that starts with the same content but is expressed differently (that is with a different expression) where as a homonym starts with the same expression but ends with a different content.

make an anatomical model for semiology that allows it to be flexible enough to expand across different fields.

Riemannian Harmony and The HarmoRubette Software

Summary. In music, the production of meaning is prominently described in harmony. This means that musical objects such as chords are given a signification as harmonic signs. The classical approach to this enterprise was proposed by Hugo Riemann (Figure 8.1).

In Riemann's function theory [60], he planned to attribute to any given chord of a musical composition one of three values: tonic (T), dominant (D), and subdominant (S). These values depend on the tonality, or *key*, in which we think in a given composition—for example C-major, B minor...

Riemann's revolutionary idea was that tonality is not described, but really *defined* by the values T, D or S, that any chord might be given. His planned to define such a function for every possible chord. He however never realized this idea for all chords, only simple triads (three-note chords, see Figure 8.2) and a number of tetrads (four-note chords) that could be dealt with.

Fig. 8.1: Hugo Riemann (1849-1919).

In Figure 8.2 we show the standard values of the seven degree triads in C-major, together with the names of the root notes.

In B-flat major, the Riemann function of a triad chord $\{Bb, D, F\}$ is a tonic (T). The Riemann function of $\{F, A, C\}$ is a dominant (D) and the function of $\{Eb, G, Bb\}$ is subdominant (S). If the tonality changes, the same chord will have a different function. For example, the function of $\{F, A, C\}$ in the key of F-major would be tonic. Harmony deals with attributing to all chords of a composition such functions.

In general, this idea is difficult to realize for complex chords. But it is the first approach to create meaning of musical compositions. In the twentieth century, several software programs have been created to calculate such Rie-

Fig. 8.2: The standard values of the seven degree triads in C-major (also denoted by $I, ii, iii, IV, V, vi, vii$, capital letters for major, lower case letters for minor or dimished triads). The English names of the root notes as notes in the context of the scale are: c = tonic, d = supertonic, e = mediant, f = subdominant, g = dominant, a = submediant, b = leading tone.

mann functions. One example is the component HarmoRubette in the Rubato software.[1]

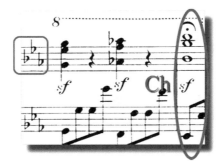

Fig. 8.3: An $E\flat$ major chord from the allegro movement of Beethoven's *Hammerklavier Sonata* Opus 106, (mm. 130-131). In this key signature, its chord's function is tonic.

8.1 The Semiotic Structure of Music

This process relies on the semiotic structure of music. First, a chord is written as an expression in the score. For example, the circled set of notes $\{E\flat, G, B\flat\}$ in our example of Figure 8.3 is such an expression. Second, in Riemann theory, this expression is given a content. In our example, Riemann would say that, in $E\flat$-major, the function of $\{E\flat, G, B\flat\}$ is tonic. The general formula for the function is given by

$$RiemFun_{tonality}(Chord) = function.$$

Referring to the chord $\{E\flat, G, B\flat\}$, its function will change if we change the tonality, for example:

---
[1] For details, see www.rubato.org.

$$RiemFun_{E\flat major}(\{E\flat, G, B\flat\}) = T,$$
$$RiemFun_{A\flat major}(\{E\flat, G, B\flat\}) = D,$$
$$RiemFun_{B\flat major}(\{E\flat, G, B\flat\}) = S.$$

The detailed calculations of the functions themselves are outside the scope of this book. You can find more in Mazzola's book *The Topos of Music* [41].

The HarmoRubette software uses Riemann functions to analyze the tonality of such a chord given the context. In a more general setup, it gives us as output values, for each chord, its functions in all tonalities. The values are however fuzzy numbers, which means that a chord may be tonic, dominant, or subdominant to a certain degree in several tonal contexts, see Figure 8.4.

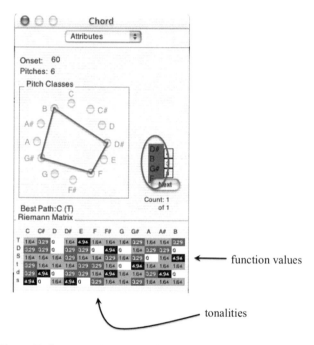

Fig. 8.4: A HarmoRubette analysis of a $G\sharp$ minor sixth chord. The function values and tonalities are labeled. Note that function values are displayed even for distant tonalities. The output values are fuzzy (continuous), rather than discrete.

Thus, the HarmoRubette software demonstrates a case in which the semiotic structure of music is exploited by computer programming. Since music follows basic principles of sign representation, it can be broken down like a language and coded into bits of information that a computer can understand. Thus, we start with an expression of written music (i.e. the score) which is transformed through the process of signification using the Riemann functions. By performing them, the computer may decipher the tonal content of a given chord. This eloquently demonstrates the semiotic structure in action.

De Saussure's Six Dichotomies

Summary. As mentioned in Chapter 2, Ferdinand de Saussure was a Swiss linguist known as one of the founding fathers of the field of semiotics. In one of his most important theories, de Saussure presented six dichotomies that describe the characteristics of signs and symbols. In this section we will discuss and provide examples for each of these six dichotomies.

9.1 Defining the Dichotomies

9.1.1 Signifier/Signified

The signifier/signified dichotomy deals with the differentiation between the expression and the content being represented. The signifier is an expression (e.g. musical notes in a score) that represents a certain meaning (e.g. pitch or rhythm). This is captured by the process of signification described in previous chapters, in which a sign goes through the stages of Expression, Signification, and Content. Music is often expressed by the surface of a score. This surface is full of expressions in the form of dynamics, written directions, and notation. Each of these are expressions (signifiers) that represent a musical meaning (the signified) such as rhythm, style, or pitch. There is some ambiguity as to where an expression ends and becomes meaning. In Chapter 10 we will discuss this in the context of more complex sign systems.

9.1.2 Arbitrary/Motivated

The arbitrary/motivated dichotomy has to do with the process of creating meaning. More accurately, it has to do with the way in which the expression is related to the content. If the expression is not related to the content, then it is said to be arbitrary. A sign is *motivated* if its expression is connected (that is analagous) to the content. This can be demonstrated by the case of *long play* (LP) records, also known as *vinyl records* and compact discs (better known as CDs), see Figure 9.1.

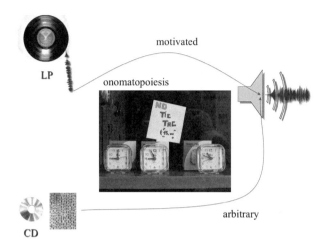

Fig. 9.1: An LP signifies its sound content in a motivated way, as opposed to a CD, whose digital expressions are arbitrary.

The grooves in an LP disc are analogous to the musical sound (variable air pressure) that is produced. This is due to the fact that an electro-magnet in the head of the needle moves up and down in correspondence with peaks and valleys of the groove. This is translated into fluctuations in electrical current. These fluctuations cause a cone in the loudspeaker to vibrate, creating sound. This process is motivated, because the grooves in the LP disc directly influence the way in which the sound is created.

The surface of the CD is covered in bumps, which reflect light from a laser in the disc drive. This reflection contains binary information that is perceived by the laser lens, and decoded by the disc-driver or computer. The information is written in *binary*, a code consisting of 1's and 0's, in which 1 means "present" and 0 means "not present". Unlike the micro-grooves of the LP disc, which move up and down in correspondence with the encoded music, the binary information on the CD does not necessarily mirror the movement of the sound.

9.1.3 The Digital Approach, Sampling

Philips and Sony have used digital encoding since 1982 upon recommendation by Herbert von Karajan. It is remarkable that his acceptance of the CD quality (characterized by a 20 kHz upper frequency limit in the Fourier spectrum) was decisive for these companies, although Karajan's age (around 50 then) was not

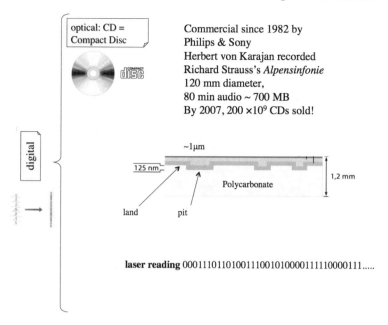

Fig. 9.2: The hardware display of a CD. The CD's laser reads the land and pit levels engrained upon the polycarbonate carrier as a bit sequence of zeros and ones.

ideal as a reference for faithful sound perception. The hardware display of a CD is shown in Figure 9.2. The CD's laser reads the land and pit levels engraved upon the polycarbonate carrier as bit sequence of zeros and ones.

The transformation from the analog soundwave to the digital representation on the CD is explained in Figure 9.3. The wave is quantized in two ways: First, the sound's amplitude is quantized by 16 Bits[1]. This means that we are given $2^{16} = 65,536$ values $(0, 1, 2, \ldots 65, 535)$ defined by the binary integer representation $b_{15}2^{15} + b_{14}2^{14} + \ldots b_1 2^1 + b_0 2^0$, $b_i = 0, 1$. Second, the (quantized) values of the wave are only taken every $1/44, 100$th of a second, i.e. the sample rate is $44, 100$ samples per second. In total, this gives 635 MByte(1 Byte = 8 Bit) CD capacity for one hour stereo recording. This allows for partials up to frequency $44, 100/2 = 22.05$ kHz. The human ear is known to perceive up to 20 kHz.

Karajan, in his fifties, could very probably not hear more than 15 kHz, so for him the quality of the CD was perfect. A young human however would have asked for higher resolution. In recent technology, a sample rate of 96 kHz with amplitude quantization of 24 Bit is being envisaged.

This looks like the endpoint of a long development of sound conservation and transfer, for which the 200 billion CDs sold by 2007 is a good argument. However, the creative argument comes from the critical concept: the container

---

[1] A one-Bit quantization would allow for just two values, 0 or 1.

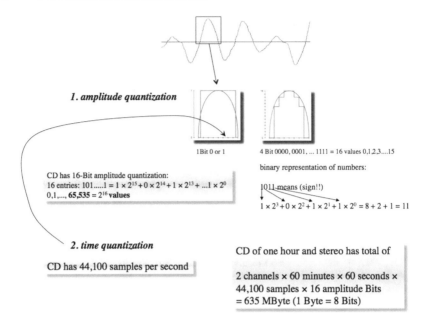

Fig. 9.3: The transformation from the analog soundwave to the digital representation on the CD.

unit of music, the CD. What is a wall thereof? First, its material part, this disc, why should this be the container? And second, why should music be transferred using such a hardware container?

Because the internet has defined the global reality of information transfer since the early 1990s, the answer to the above questions has become straightforward: Of course music data transfer can be accomplished via Internet.

9.1.4 Syntagm/Paradigm

Saussure's syntagm/paradigm dichotomy takes a central role in providing context to the semiotics of music. This dichotomy is based on two axes of meaning: the syntagmatic and the paradigmatic. The syntagmatic axis describes the position and placement of symbols (particularly their placement in time), while the paradigmatic axis concerns groups of signs with similar meanings (called associative fields). The syntagmatic axis usually deals with tempi, rhythms, and note order while the paradigmatic axis deals with the use of similar melodic patterns to create or resolve tension. In his work [27], Roman Jakobson suggested that the projection of the pragmatic axis onto the syntagmatic axis is what gives poetical depth to symbolic meaning. He called this process of (communication, art, language) the *poetic function*. The poetic function of communicative symbols is based in intentionality, and focuses on the importance of choosing

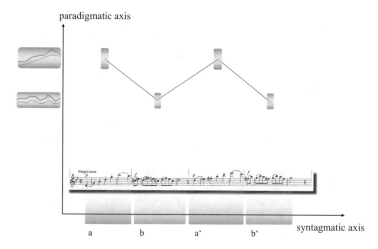

Fig. 9.4: Associative fields in their syntagmatic juxtaposition in a musical sentence.

signs carefully. Figure 9.4 shows the graphical position of associative fields in their syntagmatic juxtaposition in a musical sentence.

9.1.5 Speech/Language

The Saussurean dichotomy of language/speech adds greater diversity and complexity to the layers of musical meaning. Language refers to the system of signs that is used with common rules across all users. In music, this is musical notation. Speech refers to the way in which an individual uses language in a characteristic way. In music, this is representative of performance, or the music itself.

The conductor is given this creative license—to manipulate the group to ensure that a common language is understood throughout the whole ensemble. The language of the group, however, may be semi-dependent on the conductor's interpretation of the music. That is, the conductor's speech, his or her individual patterns of musicality, will influence what the ensemble must accept as a common language. The conductor, in this way, is able to creatively express his or her musicality through the ensemble. Despite this, each player still adheres to their own level of speech in that no other musician can play in the exact same way.

Soloists in particular are given an opportunity to express their musical speech, even to the extent of disregarding the conductor. Despite being most prominently observed in the conductor and soloists, all performers in the ensemble possess their own musical speech. It is the combination of these unique musical identities playing in concert with one another that gives rise to the incredible diversity of different, distinct ensembles and performances, even given the same instruments and vocal parts. It is this speech aspect of music that makes it at all possible to produce creative expression of individuality through music. Understanding the language/speech dichotomy in music is crucial for the creation of a meaningful interpretation.

9.1.6 Synchrony/Diachrony

Saussure suggested that a semiotic system involves symbols that are distributed in both time and space (Figure 9.5). To account for this, Saussure proposed the synchrony/diachrony dichotomy, which acts as a sort of coordinate system. Here we will use it to identify music by its time and place in history.

The synchronic axis focuses on "place" and represents what is happening music across all forms at any one time. The units of this axis are categorical, focusing on classifications of musical form such as culture or genre. It deals with music ethnology or ethnography. The diachronic axis describes the development of a form of music over time. The units of the diachronic axis are measured in time (usually in years). Typical structures of diachronicity are the history of words, i.e., their etymology.

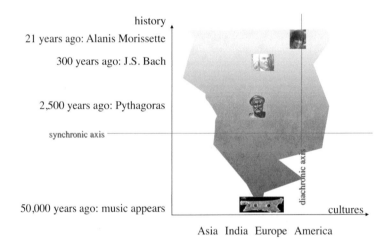

Fig. 9.5: The diachronic and synchronic axes of music semiotics.

9.1.7 Lexem/Shifter

For every word there is an objective definition. However, the meaning of a word may change due to context. For example, if I say "you" it has a different meaning than when you say "you". When I say "you", I am referring to you, whereas when you say "you", you are referring to me. This rather confusing conundrum is the concept from which the lexem/shifter dichotomy is born.

Lexem refers to signs that can be objectively defined, while shifters are signs that change meaning depending on the context of their users. In music, the majority of symbols can be considered shifters. This makes interactions with music highly dependent on context and interpretation.

Since the composer cannot write every intention into the score, he or she relies on the performer to bring out the meaning that cannot be derived from the symbols on the page. It is the task of the musician to add a human element, an element of creativity, as an interpretation of the score. This individuality is expressed through speech (as discussed previously). See Section 9.3 for an example.

9.2 Speech and Language Examples: Bach and Schönberg

In a semiotic system there are over-arching rules which define the social usage of the system. For example, grammar in a language system. There is also an individual usage of a semiotic system which may deviate from the over-arching rules. For example, in the language system it is the individual composition of sentences. Or in gastronomy you have a recipe which is part of the system, but the re-creation of the dish itself is an individual expression.

Fig. 9.6: Johann Sebastian Bach (1685-1750).

In music, the language perspective is typically shared by reference to eternal values such as Pythagorean principles or the compositions by Johann Sebastian Bach (Figure 9.6). As a devout Christian, Bach's compositions reflect a desire to represent the beauty of an eternal system created by perfect deity. His music is known for immaculately following the rules of music theory such as counterpoint. This adherence to pre-defiend rules makes Bach and his compositions a perfect example of the concept of system.

In opposition to this eternal system, Arnold Schönberg (Figure 9.7) invented an individual human approach to composition, the twelve-tone method, in 1923. Schönberg's music is largely atonal, and often not especially pleasant, especially for those who are used to listening to consonant chords. Schönberg's divergence from the normative standards is an excellent example of Saussure's concept of speech. Schönberg, unsatisfied with the pre-existing rules of music theory, created his own language in which to express himself as a creator of new rules.

Fig. 9.7: Arnold Schönberg (1874-1951).

Music in the twelve-tone scale has received mixed feedback. Many people simply find the music unpleasant to listen to, an attitude that was felt strongly by many critics as well. Yet, others supported Schönberg, considering his new system to be a work of genius. You can decide for yourself whether or not Schönberg's music is good, but for the purposes of this book it is important to note that his creation of an entirely new composition style is emblematic of the concept of speech.

9.3 Semiotics in Music Performance: the Example of Celibidache's Ideas

Fig. 9.8: Sergiu Celibidache (1912-1996).

For this section of the book, we would like to recommend that you go to YouTube and search for Sergiu Celibidache (Figure 9.8) conducting Gabriel Fauré's *Requiem* with the London Symphony Orchestra. The videos not only exemplify the concepts of semiology that we have heretofore discussed, they also provide unique insights and explanations from one of the most famous conductors in European history.

Sergiu Celibidache can certainly be called a man who was stuck in his ways. Celibidache considered music as a form of spirituality, and was quite stringent in his reverence for it. In fact, he generally would not allow his music to be recorded, as he believed this removed the authenticity of it. In his own words, Celibidache considered recordings "a standardization, a dehumanization of the reaction [music]". He was extremely vocal about the importance of music as an expression of individuality, as well as an excellent director who was able to communicate to all of the members of an ensemble in order to coordinate the magnificent power of a symphony.

As you may have noticed already in this short explanation alone, Celibidache's assertions are highly reminiscent of the Saussurean dichotomies. In particular, Celibidache seems to emphasize the importance of the language/speech and lexem/shifter dichotomies. He focuses on how music is simultaneously an expression of individuality (speech) and a comprehensive group effort to interpret the piece (language). In reference to the lexem/shifter dichotomy, Celibidache emphasizes the importance of the "shifter" quality in music performance. In fact, Celibidache suggests that music performance has a shifting nature.

Let us begin with the language/speech dichotomy. As you remember, language refers to the system of symbols that is used with common rules across all users. In music, this is staff notation. Speech refers to the way in which an individual uses language in a characteristic way. In music, this is representative of individual interpretations, performances, and the way in which the music is realized. In the videos, Celibidache often stops his ensemble to correct articulation or pronunciation. For example, in Faure's *Requiem*, he stopped his ensemble for their pronunciation and articulation of the phrase "Sanctus Dominus", saying "Not on the spot, please. Sanctus DoOMinus, in-between, not on DOMinus." In this way Celibidache was working on establishing common language rules for his choir, ensuring that the collective group was performing in the same, desirable way.

In an ensemble, the conductor is has such a creative license—to guide the group's performance to ensure that a common language is understood throughout the whole ensemble. The language of the group, however, is usually dependent on the conductor's interpretation of the music. Thus, the conductor's speech, his or her individual patterns of musicality, will influence what the ensemble must accept as a common language. The conductor, in this way, is able to creatively express his or her musicality through the ensemble. Despite this, each musician still adheres to their own level of speech in that no other musician can play in the exact same way.

However, there are some special cases in which the speech of an individual is allowed to override the language of the entire group. Soloists, for example, are given the freedom to express their musical speech, even to the extent of leading the conductor. In explanation of a soloist's opportunity for self-expression, Celibidache said "Away from your possibilities, not out—not for me—but for yourself is a little bit better" thus explaining to the soloist that he was free to express some artistic interpretations.

To the soprano soloist in the Fauré's *Requiem* Celibidache said "You are the queen, we have to bow. The question is not to be on time, because the time is not there. You create the time." In this example, Celibidache illustrates the power of musical speech, the freedom to express with your own character whatever artistic interpretation of the music that you hold.

Despite being most prominently observed in the conductor and soloists, all performers in the ensemble possess their own musical speech. It is the combination of these unique musical identities playing in concert with one another that gives rise to the incredible diversity of different, distinct ensembles and performances, even given the same instruments and vocal parts. Through speech, it is possible to produce individual creative expression. Understanding the language/speech dichotomy in music is crucial for the creation of a meaningful interpretation.

In regards to the lexem/shifter, dichotomy of music, Celibidache emphasizes the importance of live music. In the semiotics of music, the majority of signs can be considered shifters by nature. This makes interactions with music highly dependent on context and individuality. In reference to the use of a score, Celibidache says "This is why he wrote the text—which is unreliable and uncomplete—we must find what is not written in the score." To Celibidache, the importance of the shifter state of music is quite clear. Since the composer cannot write every intention into the score, he or she relies on the performer to bring out the meaning that cannot be derived from the symbols on the page.

In addition to the interpretative dimension added by the performer, Celibidache asserts that the meaning of music is affected by the performance itself, including the tempo, acoustics, and location. For this reason he never openly condoned the act of recording musical performances. "The reason I never made a recording is that I never found anyone who could make a recording. Those are photos! Photos of a reality that cannot be photographed. Do you play the record in the acoustics where they have been taken? The acoustics of a concert hall has such an importance—the tempo depends on it!" Celibidache continues to argue that the process of recording is a dehumanization of the music, taking away much of the character that the musicians work to create. To Celibidache, musical creativity not only depends on the importance of understanding the separation of music from musical notation, but also the understanding that music is shaped largely by its context. Without a contextual setting, we cannot experience music in the same way, it loses a part of its meaning.

10

The Babushka Principle in Semiotics: Connotation, Motivation, and Metatheory

Summary. In this chapter, we introduce the idea of what we call the *semiotic Babushka Principle* (sign systems within sign systems) presented by Louis Hjelmslev. We discuss the implications of applying this principle to the analysis of music and musical scores. We argue that, through the conceivably infinite mapping of connotational systems, music is capable of accomplishing significant symbolic depth.

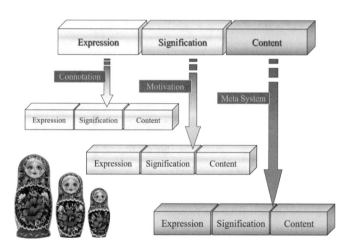

Fig. 10.1: Hjelmslev's ramifications of a semiotic system into connotation, motivation, and metasystem.

As we have seen in Chapter 6, expression, signification, and content are the basic components of meaningful signs and symbols. However, it is somewhat

© Springer International Publishing AG 2016   81
G. Mazzola et al., *All About Music*, Computational Music Science,
DOI 10.1007/978-3-319-47334-5_10

ambiguous which part of a signs corresponds to which step in the process. For example (Figure 10.2), you might say that the expression of the word "firework" is the written word itself, the signification being the understanding of the word, and the content being noise, light, and color. However, your friend may suggest that the noise and color associated with firework are the expression, where as a connotative meaning such as "Fourth of July" may make up the content.

In reality, neither of you are wrong. Rather, you are operating on different levels of symbolism. The role of connotation, brought up by your hypothetical friend (we realize that your friends probably don't casually have conversations about semiotics) was actually first brought up by Louis Hjelmslev. Hjelmslev's Babushka Principle is the idea that, through connotation, one part of a sign system can become a sign system of its own, resulting in a possibly infinite system of signs systems within sign systems.

The Babushka Principle is exemplified by the expansion of the original three components to a sign system—expression, signification, and content (see Figure 10.1 for a visual representation). An expansion of the expressive dimension is called *connotation,* while expansion in the signification dimension is called *motivation* and in the dimension of content is called *meta system.* Figure 10.1 shows these ramifications, and Figure 10.2 shows the connotational double articulation in language.

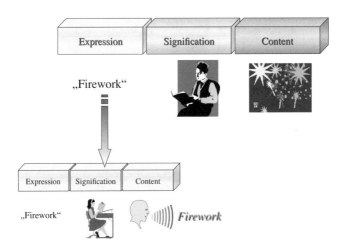

Fig. 10.2: Double articulation in language, a connotative structure. It extends Figure 6.1.

For example, let us look at how a musician perceives, analyzes, and performs a score, see Figure 10.3. The first sign system that we may conceive is that consisting of the score's text, the reading of the score, and the conception of the score by the composer (expression, signification, and content respectively). However, this entire sign system only accounts for the perception of music, ignoring the importance of analysis and performance. The semiotic nature of music does not end there, but the first sign system in itself is complete.

In order to solve this issue, we must also be aware of a second sign system involved with score analysis, including the conception of the score, the musician's analysis, and the form the music takes in the composer's mind after analyzing it (again, expression, signification, and content respectively). This sign system is a connotational system of the original sign system. That means that it takes the content of the first system (i.e. the conceived score) as an expressive level, and expands it into a sign system of its own. Similarly, we can say that the original sign system is a connotation of the second sign system, because it takes the expression (i.e. the conceived score) and expands it into a sign system of its own.

This process of condensing and expanding sign systems is responsible for the translation of the score into a performance, and ultimately to the listener's interpretation of the performance (see Figure 10.3). One can even argue that sign systems underly the way in which a society views a piece of music or a certain performer. Such an assertion is reminiscent of the psychological concept of schema formation, in which individuals form an understanding of something new through activating an intricate, underlying web of connections of related concepts. The convergence of these theories lends support to their validity.

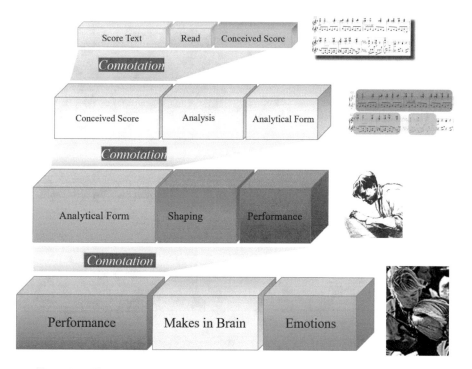

Fig. 10.3: The multiply connotational imbrication of a musical performance.

Part IV

Communication

What Is Art?

Summary. In this chapter, we will begin to define art. In particular, we will focus on defining music as a form of art. To do this, we will first consider the contributions of artists, thinkers, and poets such as John Cage, Karlheinz Stockhausen, Jean Molino, and François Villon. We will then provide examples from many musical genres, including classical music, opera and movie soundtracks, and pop music. We will end this chapter with contributions from jazz, discussing the works of Miles Davis and Guerino Mazzola's Tetrade Group.

$$-\,\flat\,-$$

Art is a way to communicate to people. Moreover, communication is also used inside art. For example, there is communication from composer to performer, from performer to performer, and from conductor to performer. According to Jean Molino, art requires a tripartition, much like communication. Communication requires a message, a sender, and a receiver. Similarly, Molino declares that the tripartition of art has a *poiesis*, a *neutral niveau* (the work), and an *aesthesis*. The parallels between these two concepts are important as we start to analyze what constitutes 'good' music.

11.1 John Cage

John Cage (1912-1992) (Figure 11.1)) is perhaps the most provocative name throughout all of avant-garde music. Whether his creations are profound works of art, or tasteless parodies of art, is highly controversial. This makes Cage and his most famous composition, *4'33"*, a worthwhile introduction to the basic definition of art.

4'33", written in 1952, lacks any musical notation whatsoever (Figure 11.2). It is arranged for any instrumentation. The only instructional note insists that the players *do not* play their instruments. The three-movement work lasts

© Springer International Publishing AG 2016
G. Mazzola et al., *All About Music*, Computational Music Science,
DOI 10.1007/978-3-319-47334-5_11

precisely four minutes and thirty-three seconds. In theory, the silence *is* the music.

In reality, there is never just silence. In a concert hall, there are hundreds of bodies breathing, coughing, and fidgeting. On stage, chairs creak beneath the musicians, and the sound of every little scuffle is projected to the very back of the hall. Even in a practice room, there is a constant whirring of the vents and the buzzing of the fluorescent lights overhead.

Fig. 11.1: John Cage in a cage (1912-1992).

This definite lack of true silence during any performance of *4'33"* was a very intentional move on Cage's part. He wanted his audiences to realize their etiquette during the performance. No matter how the audience behaved, it would be literally impossible for them to not have an impact on the performance.

Cage also sought to challenge the idea that music must be *intentional sound*. If music is fundamentally constructed of noise and the absence of noise, then he supposed that the absence of noise ought to be an accepted composition.

Fig. 11.2: The printed cover of John Cage's *4'33"*.

4'33" is either a very musical composition or a composition which causes one to think about music. However, is it art? As for the poiesis, John Cage embodies the first stage. Without him, there would be no work, and no ensemble would gather to "perform" this work (the so-called neutral niveau). There absolutely must also be an aesthesis, namely an audience to receive, and consequently realize, the work.[1]

The conflict over the artistic nature of *4'33"* lies in the fact that the creator put forth zero content. However, given that there is a work, we must also remember that it—as an individual object—will vary in time. It will never be performed in the same physical reality twice. We do not have the acoustic, much less the atomic power, to execute anything with perfect precision.

---

[1] Aesthesis refers to perception, not to be confused with aesthetics, the theory of beauty.

Therefore, we propose that Cage's work is actually an empty work, without content. He constructs the real art in the extreme form of an empty 'canvas'.

11.2 Stockhausen's *Klavierstücke*

4 5 3 1 2 : 2 1 3 5 4

Fig. 11.3: The 12-tone tone row from Stockhausen's Klavierstück IX. The numbers indicte the invervals between successive notes, showing their symmetric arrangement.

Karlheinz Stockhausen (1928-2007) was a controversial composer in the second half of the 20th century (Figure 11.4). His predecessor and renowned dodecaphonic composer, Arnold Schönberg (1874-1951), had an immense influence over the direction of his work. Many of their compositions were rooted in mathematical set theory and elementary combinatorics. Compositions of this nature may be termed *serial*. More visually, this kind of composition can be likened to a deck of cards. You can reverse the order of the cards or shuffle them into various arrangements. Imagine all aspects of an original piece of music 'adjusted' or 'remixed' to form some new combination. Volume of notes (including attack, sustain and decay), execution (including ornamentation), and intervals (in the case of chords) are just some of the musical elements that can be 'shuffled' and then notated in a score.

Serialism developed throughout Stockhausen's cluster of works titled *Klavierstück*. These works are important to our discussion, for serialism completely demolishes the connection between the audience and the creation or creator. Stockhausen, as the creator, is strongly connected to the existence of the work. However, the audience is completely disjunct from the combined object of the creation and the creator. It is impossible to guess the poietic germinal nature of the 12-tone tone row of Klavierstück IX shown in Figure 11.3. Although the creation of the tone row may have some meaning to Stockhausen, its randomized nature means nothing to an audience, halting the cycle of artistic communication.

Fig. 11.4: Karlheinz Stockhausen (1928-2007).

11.3 Molino's Tripartition and Realities.

Jean Molino is a semiologist who worked at the University of Lausanne in Switzerland. He has extensively examined the function of signs within music. His studies include the popular idea that music is a language that requires a level of mastery to be considered an *art*.

The Molino tripartition constructs a three step process of creation (Figure 11.5). In short, there is beginning which he calls the *poiesis*. This is followed by a *neutral niveau,* some sort of product, the work or composition. Finally, there is the *aesthesis,* the reception of the work by the audience that is equally as important.

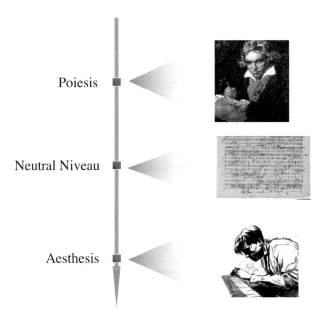

Fig. 11.5: An example of Molino's tripartition.

As it pertains to our argument, Molino defines the differences among who makes art, who receives it, and the intermediate message between them.

If we visualize artistic creation as a shape, it is best to imagine it as a set of three concentric circles, see Figure 11.6. At the very center, we find the poiesis. This point is necessary for the rest of the circle to develop. Therefore, we can say that the art is primarily dependent on this. Continuing with that premise, we infer that the creator is the most important contributor to the creation. It is essential, and actually quite defining, that artists are the very center of their work and their message. We will discuss various artists and their egos in Section 11.5.

The creation is then an extension of the creator, moving out in all directions. This is the body of the circle. The neutral niveau, the "meaningless" level, is nearly an entire circle. But without its central point, its creator, we would miss the point. Without its center and without its edge (which we will discuss momentarily), it exists in an incomprehensible infinity.

Aesthesis tames the infinite nature of a work of art. It can be represented by the circumference of a circle, binding it to a specific size. Reaching aesthesis indicates that a whole process has been undergone. This process, is the process of signification as discussed in Chapter 8. The specific size of the circle is the message as it is interpreted by an audience or viewer.

Fig. 11.6: Maggie O'Brien's visualization of the Molino tripartion.

11.4 The Babushka Principle in Communication

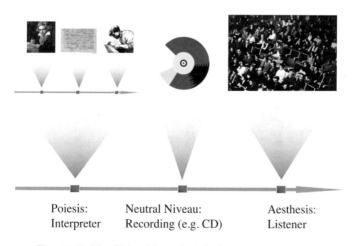

| Poiesis: | Neutral Niveau: | Aesthesis: |
| Interpreter | Recording (e.g. CD) | Listener |

Fig. 11.7: The Babushka principle for communication.

We can extend Jean Molino's framework to an almost infinite level. Subsystems of communications exist within one another (Figure 11.7). The length of analysis only depends on the time allotted to the analyst. This is the same principle which we discussed more generally within semiotics (Chapter 10).

For example, let's start with Dmitri Shostakovich's *Fifth Symphony*, opus 47. Just after he was condemned by the Soviet administration for the nature of his previous symphonies, he composed this four-movement work in the summer on 1937. His neutral niveau, the resulting score, was presented to the Leningrad Philharmonic Orchestra. The conductor Yevgeny Mravinsky, and his musicians then interpreted the score. They in turn produced a soundscape, another neutral object, which was received by the audience. Allegedly, the audience was so moved by the whole event that they were brought to tears and a lengthy ovation.

In modern times, this process is even more prolonged. The symphony hall may be rigged with microphones which receive the sound waves and interpret them as electrical impulses. These impulses can be taken into the computer, stored in ones and zeros, and then burned on to a CD. Again, the CD can be read by another laser which translates the message back into electrical impulses, which can be used by a speaker to output sound. Finally, someone listening to the CD ends this process. Their interpretation ultimately outlines the extent of the whole production. As you may now see, communication is a constant cycle within the experience of art.

11.5 Examples of the Poetic Ego in Artistic Communication

Although we can define art, it cannot be fully realized without experience. Therefore, we found it fitting that we provide you with ample references for various types of artists creating different types and qualities of art.

Fig. 11.8: The Art Ensemble of Chicago.

11.5.1 Art Ensemble of Chicago

The Art Ensemble of Chicago (Figure 11.8) is an avant-garde jazz ensemble that formed in the 1960s. While their members can be seen playing saxophones and trumpets, they do so in the midst of hundreds of miscellaneous instruments.

The members of the ensemble communicate spiritually through their multi-instrumental music. They call upon a fictitious mythology. They do not fulfill the stereotype of the "angry black jazz musicians". Instead, they're more similar to dadaistic priests. Their performances are now rituals instead of revolutionary manifests. Their entire poietic position is as central as it is different from traditional free jazz approaches.

11.5.2 Alanis Morissette

Born to military parents, Alanis Morissette (Figure 11.9) was urged to develop a solid sense of self at an early age. During her teenage years, Morisette struggled with depression and multiple eating disorders, stemming from violent private experiences. She found that music was a suitable outlet for her intense emotions. She was able to convey her violation through scathing, straightforward lyrics. This allowed the audience direct access into her mind. Fans could

Fig. 11.9: Alanis Morissette (1974-).

feel her anger, and identify with it themselves. Her unexpected success with her first single, entitled *You Oughta Know*, for the album *Jagged Little Pill*, allowed her to continue to use her image to express her anger and distress. The extent of her personality can be seen culminating at the 2004 Juno awards where she wore only a bathrobe and a flesh bodysuit. Throughout the years, she continued to make music and convey her raw emotions to the public, allowing her to become a strong public figure.

11.5.3 Angel Haze

A self-proclaimed agender pansexual, Angel Haze (Figure 11.10) utilizes her ego in order to demonstrate her special sense of freedom. Raised in a hyper-religious community, she was restrained from participating in society as a youth, leading to her lash out later as she delved into the world of music. She released several mixtapes online for free as she began her career. She was later signed by Republic Records, where she released her first album, *Dirty Gold*. A pro-

Fig. 11.10: Angel Haze (1992-).

ponent for free love and unbridled self-expression, she has become an icon of rebellion and freedom in the rap world.

11.5.4 Michael Jackson

Fig. 11.11: Michael Jackson (1958-2009).

Michael Jackson (Figure 11.11) was born into a family of musicians. Getting his career start as a member of the Jackson 5, he quickly gained popularity for both his stunning vocals and boyish charm. As he aged, his solo career took off, and Michael developed into a pop icon. Revolutionizing the music video, diversifying pop music, and perfecting the entertainment factor of stardom, he influenced an entire generation with his music as well as his humanitarianism. His ego became a placeholder for young fans, and he became a cultural phenomenon. Through his personage, he was able to connect with his audience on multiple levels. His lyricism, musical background, and physical presence and dance shaped his poetical ego. His art had a triple position in a connotation from music, expressing dance, which in turn expresses a text (of love).

11.5.5 Jackson Pollock

Described as having a volatile personality, action painter Jackson Pollock (Figure 11.12) struggled with alcoholism his entire life. He utilized his art to grasp and demonstrate his individual feeling, focusing on the physical process of creation to express himself, not the external product. His images contain living lines and energetic designs. He allowed his viewers to feel his perceived position in existence, without needing to identify what they were actually witnessing.

Fig. 11.12: Jackson Pollock (1912-1956).

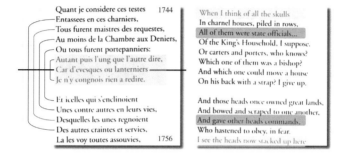

Fig. 11.13: A display of the parallels in Villon's work.

11.5.6 François Villon

François Villon (Figure 11.14) was one of the best known French poets of the Middle Ages. His piece, entitled *Le Grand Testament*, utilizes symmetric verse organization in order to shape a view of the world around his own perspective, see Figure 11.13. He addresses seeing multiple social positions all at once, and arranges these observations cyclically. This allows the reader to form their version of the world around Villon's ego, instead of objectively as is usually custom. The syntax of the poem gives it the power to organize the thoughts of the reader without directly doing so.

Fig. 11.14: François Villon (1431-1463).

11.5.7 Stolberg-Schubert

The poetical ego may also transcend different art forms. For example, Franz Schubert (1797-1819) took a poem written by Friedrich Stolberg (ca. 1750-1819) in German, entitled "Auf dem Wasser zu singen—für meine Agnes", and turned it into a musical work (Figure 11.15). Schubert utilized three-note motives and paired them with the ego of the original work, thereby mimicking the poetic

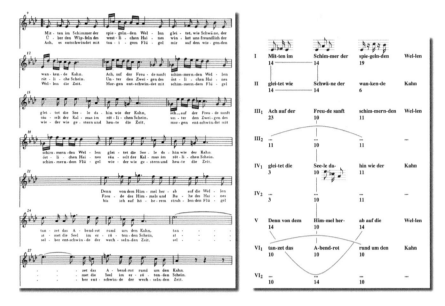

Fig. 11.15: Left: Franz Schubert's setting of the text in 1823. Right: A diagram outlining its motivic symmetry.

ego put forth by Stolberg, once again ceding the power of the work to the focus on the individual ego within it.

LIED AUF DEM WASSER ZU SINGEN, FÜR MEINE AGNES

Mitten im Schimmer der spiegelnden Wellen
Gleitet wie Schwäne der wankende Kahn;
Ach, auf der Freude sanftschimmernden Wellen
Gleitet die Seele dahin wie der Kahn;
Denn von dem Himmel herab auf die Wellen
Tanzet das Abendroth rund um den Kahn.

Über den Wipfeln des westlichen Haines
Winket uns freundlich der röthliche Schein;
Unter den Zweigen des östlichen Haines
Säuselt der Kalmus im röthlichen Schein;
Freude des Himmels und Ruhe des Haines
Athmet die Seel' im erröthenden Schein.

Ach es entschwindet mit thauigem Flügel
Mir auf den wiegenden Wellen die Zeit.
Morgen entschwinde mit schimmerndem Flügel
Wieder wie gestern und heute die Zeit,
Bis ich auf höherem stralendem Flügel
Selber entschwinde der wechselnden Zeit.

SONG TO SING ON THE WATER, FOR MY AGNES
In the midst of the shimmer of reflecting waves
The bouncing rowboat glides like swans.
Ah, over the gently shimmering waves of joy
Glides the soul like the rowboat.
From the heaven on down the waves
Dances the evening glow around the boat.

Above the top of the grove in the west
The red glow in the sky winks to us;
Below the branches of the grove in the east.
The reeds rustle in the red glow of the water.
Heavenly joy and the peace of the groves
The soul breathes in the evening glow.

Ah, time fades away from me
With wet wings on rocking waves.
Time flies on shimmering wings.
Tomorrow just like yesterday and today.
Until I myself on more highly radiant wings
Flee from the changing time.

Fig. 11.16: Raffaello's *School of Athens.*

11.5.8 Raffaello's School of Athens

A fresco created by the Renaissance painter Raffaello in 1509, *The school of Athens* shows an image of philosophers in ancient Greece centered around Platon and Aristoteles, see Figure 11.16. This painting articulates a visual representation of the human ego, addressing the central perspective (which had previously not been attempted) instead of the more divine and eternal viewpoint. Within the painting, there are two different perspectives, giving the illusion of cinematographic movement of approaching to the observer. This allows the viewer to enter into the perspective of the painting, making the ego the primary focus of the art, not solely the image.

11.6 The Opera and Music for Movies

It is a common saying that the best movie music is that which is not consciously heard. We would like to argue that in a movie it is all about music, that the image is just the surface. We first discuss the origins of music for movies, the opera, which is an early version of multimedia art. We then discuss some significant movies.

11.6.1 Opera

Why do we need to learn about opera? As with other topics, learning the past is a way to the future. However, opera is not only linked to the past. Although it has evolved with the centuries, the main masterpieces are preserved, performed, and appreciated all over the world. There are also contemporary opera composers. Opera represents the character of a composer, as well as an entire generation of society, and sometimes even humanity's spirit or universal ideas of a period.

Opera was for the past centuries what the movie is for today. With the difference that, once it is made, the movie cannot be changed, but the same opera, even based on the same score, is different *every* time it is performed. Little differences, like the timbre of voice, small tempo and intensity variations, and—of course—gestures are of capital importance in making every performance unique. In this way, the performance is a shifter as described in Section 9. In a lyric opera, a narration is developed in the music: the actors are singing. The 'magic' of opera is in this fiction and the very non-realistic way it's recited. When well performed, opera captures the attention of the audience and conveys a sense of the sublime.

Opera was born between the 17th and the 18th century. It originates from the sung sacred scenes first represented in churches, and then in public places. Unlike other events in history, we have the precise date of the first opera: on October 6, 1600, *Orfeo ed Euridice*. The story was based on Ottavio Rinuccini, and the music was composed by Giacomo Peri. Other early opera composers include Giulio Caccini, Claudio Monteverdi, and Christof Willibald Gluck.

If we want to uncover the ancient roots of drama and music, we should go back to Ancient Greece. However, we must look past the general distinction of tragedies and comedies.

In the beginning, the genre of opera was characterized by closed forms with *recitativo* and *aria*. In the recitativi the text was dominant, with slight pitch variations, more free tempo, and often a simple chord accompaniment. Recitativi are in general very hard to sing. They are then followed by arie, more expressive and music-dominant pieces. Arie are also present in more modern operas, and are often performed as isolated pieces during concerts.

As the genre evolved, recitativi mainly disappeared (and were even substituted by just spoken sections in French and German tradition). The separation between sections of the opera progressively disappeared as well, leaving more continuous musical sequences. This process led to the so-called *endless melody* of Richard Wagner's operas. A similar tendency to longer scene and musical sequences is also present in the last works of Giacomo Puccini. Another similarity between these two composers is the use of *leitmotiv*, a term in particular coined for Wagner's works: it is the attribution of a musical theme to a personage, or to a situation, or to an element that is particularly relevant for the plot. The same technique is also used in movies, television, and video games.

The vocal technique mainly used in the frame of opera is the so-called *belcanto*. This is a sophisticated use of the voice that assures a great projection of sound, vibrato, and use of the complete vocal register, with minimal effort and a general relaxation of the phonatory system. Traditionally, opera singers do not use amplification systems. In addition to fully utilizing the acoustics of the theater, opera singers typically have strong natural vocal skills which are well-trained for the profession.

Why, then, would operas be of general importance, and not only for musically trained people? Earlier we mentioned the tragedy representations in Ancient Greece. Tragedies (and comedies) were highly relevant and important to the people, not only for the sake of entertainment, but also as an educational tool. Opera is dominated by the same morals which can be found in stories and folklore throughout the world's history. Even when a personal story is described, or a particular fact, there are some ideas that can be applied to all people. The same happens for a contemporary song: even if the text talks about a love delusion of a specific person in a specific time, a similar sentiment of unhappiness can be shared by millions of people through the world.

Opera often contains emotional *topoi*, and it can also have a political content. This is, for example, the case of Giuseppe Verdi's *Nabucco*, when the choir of prisoners was used and correctly interpreted also as a metaphor for the desire of freedom from the foreign oppressor, and political unity, for Italian intellectual people at that time. Opera can also contain a critique of a certain way of thinking, a wrong mentality.

Another of Verdi's works, *La Traviata*, is the story of Violetta, a girl whose survival was supported by a baron, but fell in love with a young man named Alfredo. She changed her lifestyle completely for him. However, she had to stop her love story when Alfredo's father told her that his daughter couldn't celebrate her marriage, because the relationship of Alfredo was ruining the reputation of the entire family. Thus Violetta, already suffering from tuberculosis, was renounced from her last source of happiness. In a scene, she sings *Così alla misera ch'è un dì caduta / di più risorgere speranza è muta, / se pur benefico le indulga Iddio, / l'uomo implacabil per lei sarà*, that means "In this way to the poor that once fell, it is not allowed to rise again; even if God is merciful indulging to her, the man will have no pity for her—will be implacable," an evident critique of the morality of that time. Figure 11.17 shows the renown American soprano Anna Moffo singing in the title role.

Fig. 11.17: Anna Moffo singing Violetta's role in *La Traviata*.

Italy has a great tradition of opera composers. To mention just a few names, we have Bellini, Scarlatti, Donizetti, Verdi, Puccini, Rossini, Mascagni, and Leoncavallo. Italian language is widely used in the opera world: we can think, for example, of Mozart's Italian operas. However, other countries significantly contributed to this genre. First of all, we can think of France with Massenet, Berlioz, Bizet; German-speaking countries with Mozart, Wagner, Schönberg, Berg. A wide span of other languages (and cultures) are represented through opera, including Russian (Borodin, Mussorgsky), Czech (Janaček), and English (Purcell, Weill in the English period), just as a few examples.

Opera is still alive today. Even if new musical genres of narrative in music have appeared, such as musical theater, opera itself counts many new works every year.[2] Many contemporary composers are active in this field. To mention some of them, we can think of Sir Peter Maxwell Davies, Franco Battiato, Marco Betta, and Giovanni Sollima.

Still the classics are studied and performed due to their 'universality' of themes and richness of music. We can cite the emerging contribution of Asia, in particular South Korea, to this genre with important schools and new emerging singers.

---

[2] Maria Mannone has written a short opera, *Il coro degli animali*, about the controversial theme of animal experiments. Guerino Mazzola has also written the libretto of an opera, *Apocalypse Now—and Then*, about God's musical motivation (and failure) to create the world.

Let us now analyze some movie soundtracks.

11.6.2 Garden State

Fig. 11.18: A scene from *Garden State*.

Garden State (2004) was a romantic-comedy movie directed by and starring Zach Braff (Figure 11.18). Overall, it received moderately positive reviews upon its release. The soundtrack in this film was a compilation of pop songs by various artists. The unifying features among all of the songs were the constant steady beat and meter. This produced a mundane effect which hinted at life's constant ticking through time.

We are not going to argue that pop music is inherently bad. However, there is a trend in pop music which makes it less artistic. That is to say, there tends to be less communicative value in pop music as a genre. The constant beat, the regular chord changes, and the linear melodies are what make pop music predictable. This beneficial trait allows pop music to spread to large audiences, as more people can sing and dance along than to John Cage's silence.

11.6.3 Satyricon

Satyricon was a Latin satire written by Petronius. It documented the fictitious relationships between two friends, Encolpius and Ascyltus, and Encolpius' lover, Giton. In 1969, it was adapted into a surreal drama by Federico Fellini (Figure 11.19). Although it was a rough adaptation made from the remaining scraps of the original text, it still generally describes the dream-like mythical adventure of the three men in ancient Rome. The film was essentially nonlinear and had nine primary episodes.

The music and soundscape of the film was directed primarily by Fellini's close coworker, Nino Rota. One of the more interesting moments is the audio sequence describing an episode of Encolpius' personal story. A connection is

Fig. 11.19: A scene from Fellini's *Satyricon*, the dinner of Trimalchio.

drawn between Encolpius and the hero Theseus as Encolpius finds himself trapped in a labyrinth with the Minotaur.

While Encolpius is fumbling around corners and showing his lack of athleticism, the sound of a rushing wind follows him. More than just a dry desert breeze, it seems as if there is a gust moving throughout the entire maze. Above the sound of the gust, one can still hear Encolpius' gasps for air and his lame prods at the Minotaur. Suddenly interrupting the heterogeneous flow of the wind, a terse chant is heard all around. The camera pans, and it can be seen that there are crowds of spectators watching expectantly from a distant cliff overlooking the maze. Intermittently, they stop and start chanting several times throughout the chase.

This soundscape is an interesting choice. When well executed, an orchestral score can provide a great deal of drama without being noticeable and cheesy. However, Rota opted for a more unlikely approach, contributing to the surrealism of the production at large. In the film world, both the chanting and the wind effects are considered diegetic, as we can identify the source of the sound from within the film. However, perhaps due to the era in which the film was produced, there is a non-diegetic nature to the soundscape. Because it is flat, there is no sense of depth produced by the monophonic output. This creates an interesting juxtaposition between the perspectives of the characters, the filmmakers, and the viewers.

The soundtrack of *Satyricon* is not just effective avant-garde background music. This soundtrack is the soul of the film. Take it away, and you have nothing but a few loud breaths from Encolpius. The soundscape shapes the story and is the only part of the film that really makes any sense at all.

11.6.4 Fellini's $8\frac{1}{2}$

Federico Fellini's 1963 drama, $8\frac{1}{2}$ (also *Otto e Mezzo*), is all about the music (Figure 11.20). Like *Satyricon*, its soundtrack was mostly composed by Nino Rota. Through music, the entire film represents Fellini's ego as an artist. The plot features a film director (Fellini's alter ego) who has a crisis of creativity and is unable to come up with a complete new story.

Journalists, reporters, and critics pressure him to give some information about his new work, so he crawls underneath a table and shoots himself. However, some sort of twist in time and space occurs, and the movie continues as if it hadn't happened. Perhaps it was real, or perhaps it was a dream. Nonetheless, dimensions of mental and physical realities intermingle, and the dream meets the truth. Finally, the protagonist makes the movie with all the people of his life and his entire story. The narration dissolves into a great dancing scene. The people and the music vanish little by little. Eventually, only a young boy playing the flute remains: the image of the inner ego of the director. At the end, there is only the sound of the flute and darkness.

Fig. 11.20: Poster of Fellini's $8\frac{1}{2}$.

Since the time of its release, $8\frac{1}{2}$ has been considered by both, public and critics, as 'the' movie. It was universally acclaimed due to its artistic insight and its truth to the various realities of Fellini's life. His ego was prevalent in that it starred memories from childhood, but not in a narrative way as in *Amarcord*, another of Fellini's movies.

11.6.5 Onibaba

Fig. 11.21: A scene from *Onibaba*.

Sound is an essential aspect to music. The vibration of molecules in wave patterns is unique to music, as opposed to the visual arts, dance, and literature.

The film *Onibaba*, which was produced by Kaneto Shindo in 1964, is a Japanese horror film set in the fourteenth century (Figure 11.21). The title can literally be translated to *demon hag* or *demon grandma*. In the film, an old women and her daughter brutally hunt many samurais and infantry. At each hunt scene, there is an ominous "silence". However, the careful ear would recognize that this is not an empty silence. The producers have filled the scene with ambient sounds. The air can be heard harshly whipping through the tall grasses through which they stalk their victim. When they strike, a thunderous beating of drums explodes.

These scenes would not have the importance that they bear if they did not have the sounds. The sudden switch between a lack of sound and overwhelming sound draws the attention of the audience. The music is mirrored in the evil women, and vice versa. Here again, the music is the lifeline of the film.

11.7 The Infinite Production

The creative part of a musical production of a CD doesn't end with the recording. Even the recording can be dramatically changed. For example, the types, numbers, and the positions of microphones may be changed. Often, since different tracks are recorded of each musician, mixing and layering the tracks can emphasize, correct, and/or ameliorate the sound of the performance. During the so-called *post-production*, it is possible to edit, to cut, and to work on music. For example, if different cameras are video-recording the performance, the way of selecting photographs, perspectives, and juxtaposing different images can make emphasis in a detail of the performance, increasing its expressivity. All these phases can also require a certain amount of taste and creativity.

We could say that, in 'the future of recording,' the user could get the full recording spectrum and shape the perspective they want. For example, one could change the music as it can be heard by walking through the orchestra. We could even ask if there is no definitive audio file. This futuristic idea can be related to the potentiality of virtual reality and active participation of the public in the artwork.

One could ask if such possibilities implies the death of the authorship, and the negation of the objectivity of the artwork. Our position here is that we don't want to neglect the ontology of the artwork and the importance of the identity of the artist; however, we want to stress the importance of the different perspective on the same artistic *object*. In fact, the identity of the artwork is not denied, but, in the opposite way, more focused by the collection of its different perspectives. Perhaps with virtual reality the spectator will no more longer be passive, but instead become a more and more active part of the artistic process of creation.

A big open question arises while talking about user-friendly devices. Smartphones are easier to use than a terminal command, because they are gestural and intuitive. However, not everybody has the necessary technical competence to understand what's happening inside a mobile device. We don't know whether the future of music will involve more and more user-friendly devices or if it must first be extremely complex.

11.7.1 Miles Davis' *Bitches Brew*

Miles Davis' album, *Bitches Brew*, was one of the most influential works in rock jazz (Figure 11.22). The first famous musical example of infinite production, it sold 500,000 copies. It was a complete recording of the sounds produced in the studio room. This album is unique because the producer made choices in order to creatively break down the walls of traditional recording. With this album, Davis initiated the change towards a more inclusive musical culture.

This, in turn, allowed the artists to break down the walls in their creative processes. By blurring the line between the composer, musicians, and audience, artists and producers can create an interactive type of music. This allows the listeners to experience music in a new 'live' way. The blurring of these lines is a chance for the music industry and culture to blend the styles of reproductive classical and creative improvisational performance. In this new format both, the composer and the performer have the power to be creative.

The idea that the social constructs surrounding music are evolving frustrates people who believe in a "museum culture" of music in which the way that we present and interpret music must remain constant. This museum culture of music is often propagated by music conservatories, which advocate for the preservation of music style and often condemn creative progress.

However, art is an ever-changing field that finds new life with each production. If artists are "too respectful" to not give a new way of performance to the earlier compositions, the "classics" are dead. Perhaps we can also find a way to rediscover pre-existing music with new technology. This new approach allows non-musicians to experiment with the music creating process by using MIDI to make these changes (see Chapter 12.1).

Fig. 11.22: *Bitches Brew* by Miles Davis.

11.7.2 Mazzola's Tetrade Group Recording

In this section, we will discuss Mazzola's Tetrade Group, constituted by Guerino Mazzola (piano), Jeff Kaiser (trumpet), Sirone (bass), and Heinz Geisser (percussion). The interest in discussing this example is also in the context of artistic creativity: technology opens the walls of a fixed, static recording.

In spite of the mixing technique that they use, Mazzola's group tries to make the recording sound as real as a live performance.

This ideal of the true thing is sometimes also portrayed by recording labels, such as Bob Rusch's CIMP, which proudly claims in their mission statement that "What you hear is exactly what was played. Real musicians in a real space." But there is no such a thing like "the real space." The space of listening is open, infinite and, moreover, not limited to the technical equipment. By this we mean that the variety of recording spots do not define the end of listening. In addition, it is in many ways evident that *listening is a highly creative activity*, that you have to actively "ride the horse", and that in so doing, you will ride it in a new way each time.

Making their music something that is experienced in variety of ways is what Mazzola's group drives for. The idea is to use technology not to create a 'false' artistic object, instead, to recreate a more 'natural' effect, even if it could appear at a first sight as quite paradoxically. We can think of the movement of the eye looking at a musical performance: it is jumping from a performer to another, from one hand to another, and so on. With technology, for example collecting the shots from several cameras, it is possible to 'recreate' this collection of eye movements. The same thing can happen for hearing. Moving around a sound source, or listening to a moving sound source, or both, corresponds to different audio feelings. The possibilities opened by the mixing of sound tracks obtained by different microphones can generate such a 'natural' effect. Moreover, returning to the visual example, to better understand the geometry, the shape of an object, it is convenient to watch it from different perspectives. The 'truth' about the object is given by the collection of such perspectives. In music, the perspectives can be the different performances of the same piece. And also, for the same performance, and in particular for improvised music, the collection of the different recording-tracks as discussed.

The recorded variety of tracks must be shaped for the final stereo output, and this is not simply reproductive at all. Of course, a two-channel recording simplifies the post-production, but it only hides the mysteries and difficulties of distributing and balancing those microphones. Therefore, even though the music that Mazzola's group produced is not truly real, they use a sound engineer to sound as real as possible as if someone is listening to a live performance. Mazzola and his group spend as much time in the mixing process as in the recording process because they want the audience to hear the music in their own way. The open space of infinite listening suggested by a multitrack recording makes clear that the myth is not reality. Let us give an example from a recording project, which was also a unique experience for the students of Mazzola's course *Free Jazz—From Structure to Gesture* at the University of Minnesota School of Music. On February 19, 2008, the Tetrade quartet (see Figure 11.23) recorded the CD *Liquid Bridges* at the Wild Sound Studio in Minneapolis, with the sound engineer Matthew Zimmerman and his collaborator Gérard Boissy. The students attended the entire recording process of that day (from 9 AM to 6 PM) and could see the difference between "HiFi" and "HiVi", the *High Vision* music, where no fidelity to given facts is at stake. Here is the precise list of microphones:

Fig. 11.23: The Tetrade group with Jeff Kaiser (right) on electronically extended trumpet, Guerino Mazzola on grand piano (second from right), Sirone on bass (left), and Heinz Geisser on percussion (second from left).

1. Kick: EV RE-20
2. Kick 2: Shure SM91 on floor in front of kick drum
3. Snare: Shure Beta 57 top, Sennheiser 901 bottom
4. Tom: Sennheiser 901 (top and bottom)
5. Floor Tom: Sennheiser 901 (top and bottom)
6. Hi Hat: AKG C451
7. Drum Overhead: AKG C34 (Decoded in M-S (mid-side) with channel 1 cardioid, channel 2 figure 8, Matthew used the Waves S148 matrix plugin to decode, inserted in channel insert)
8. Drum Room, behind kit: Blue Dragonfly, into Distressor
9. Bass DI: Avalon
10. Bass Room: RCA 44 into Wes Dooley TRP
11. Bass Close: Neumann U87 into Manley Voxbox
12. Piano X/Y: AKG C24 into Great River MP2NV
13. Piano NR: Neumann KM184 pair into Millennia HDV3C
14. Trumpet: Blue Kiwi into Manley Voxbox
15. Trumpet Electronics: Direct line level

The dramatic difference between this multitrack and the two-channel method is that, for example, Jeff's output was twofold: directly acoustic and through the computer system. Jeff stressed that Mazzola's group should never use both outputs at the same time. It is an example of an important decision for the final artistic result, that has been postponed to the post-production step, during the phase of mixing and editing the CD on May 23/24, 2008. Moreover, the post-production was also a continuation of the creative listening process because the group had to mix the multiple inputs and to *create what nobody had ever heard*. So the listening was prolonged after the recording, and in an ideal world, Mazzola's group would just hand out to the public all tracks of an infinity of microphones and let them mix whatever they want to hear.

One word about the production conditions: The musicians must take care of being able to hear everything they possibly can. This works as follows: First, the musicians have to be separated acoustically from each other in order to obtain pure instrumental input. Then, parallel to the separation, the visual contact must be optimized since seeing the fellow musician's gestures, the bodily and facial expressions, is extremely important for the gestural communication even if all eyes are closed during the performance itself. Third, the individual headphone settings must be adjusted so that every musician can hear most transparently his and his fellows' output. This is another creative extension and shaping of the listening space, and it may be very different from what at the end will be heard on the CD, but it is crucial for the perception and interaction of sound reality.

So it was clear to the Tetrade group that the mixing and editing was just one more step in an infinite chain of creative listening, limited by nothing except the available technology. Post-production is one privileged way of listening. Let us terminate this short digression by a remark on the collaborative art's characteristics: space, gesture, and flow. The space, as already discussed, is opened by the multiple perspectives of sonic tracks. The gestural component is beautifully illustrated in the second piece "Liquid Bridges" of the CD: At 5:48, everything breaks down to a quasi-silence, where only the trumpet is quietly blowing air, no pitch, just air current. The music has come to a zero point, where time is the only event perceived and created. At that moment, the piece could have ended if the group had understood the moment as a giving away of time, as an ending of the group's creation of time. But the very quiet piano chords that follow show that the group have complete control over time; the sound reminds us of the stroke of the clock at midnight. There was time, there was all the time on Earth to go through that zero moment. This gestural relaxation is the opposite of emptiness. It is the pure gesture of the making of time that happens in this magical moment of the recording. The development of the piece to its end proves that the group has full control over the unfolding dynamics. The piece grows into an intense flowing energy and takes the listener into a crescendo of a quasi-hypnotic drive. This is at least what the sound engineer reported after listening to the final mix.

The pre- and post-production are natural and essential parts of the infinite process of the making. The technical post-production is not only a trace of the infinitely open space of microphonic ears. It in some sense represents the listening process itself, enhanced by technology. It is just the first HiVi step in the infinite listening, that is a never ending process which is unconcerned with the idea of perfection. Instead, it will exist as an autonomous gestural vibration in the making.

12

The MIDI Code

Summary. MIDI is a language for digital communication between electronic instruments and computers. It is modeled to imitate the movement of the fingers, staying down, going up and so on. It is a low level language which tells you what to do without any deeper understanding. It is very important for electronic and pop music.

$$-\ \oint\ -$$

12.1 A Short History of MIDI

MIDI (Musical Instrument Digital Interface) was officially introduced in January 1983 at the NAMM (National Association of Music Merchants) conference with its "MIDI 1.0 Specification."[1] It was created by a pragmatic community of music merchants, and replaced the first sketches of the Universal Synthesizer Interface from 1981 which was created by the academic Audio Engineering Society. No musicians or music theorists were involved in this development, they later complained about the MIDI's deficiencies. Standardization of music formats had never been a topic of music theory before mathematical and computational music theory started developing universal standards.

In the following sections we describe MIDI communication and the structure of MIDI messages. These messages are exchanged between any two computers and/or synthesizers. The general functionality of MIDI is shown in Figure 12.1. As we have discussed, music-making is a three-step process: We usually have (1) a score whose frozen gestures are being "thawed" to (2) gestures, which act on an instrumental interface and thereby produce (3) sound events. The gestural interaction with an instrument (process from the left top position to the right top position in Figure 12.1) is where MIDI's main functionality lies. The

---

[1] MIDI specifications can be looked up on the MIDI association page https://www.midi.org/.

© Springer International Publishing AG 2016
G. Mazzola et al., *All About Music*, Computational Music Science,
DOI 10.1007/978-3-319-47334-5_12

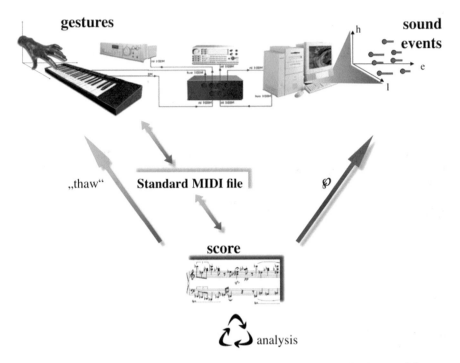

Fig. 12.1: Making music is a threefold communication: We usually have a (1) score whose frozen gestures are being "thawed" to (2) gestures, which act on an instrumental interface and thereby produce (3) sound events.

movements of the human limbs (hands for keyboard players) are encoded and then communicated to a synthesizer that produces corresponding sound events. Therefore, MIDI is essentially a simplified code for human gestures interfacing with an instrument. A second functionality of MIDI is the transformation of instrumental gestures to a Standard MIDI File such that the performance can be replayed later. Standard MIDI files can also be transformed into a digital score, and vice versa.

MIDI just tells an agent to do simple gestures at a defined time and with a specific key and instrument. It is not a coincidence that machines that understand the MIDI messages are called slaves in MIDI jargon. We can draw a similarity between a conductor and his orchestra. The maestro waves his arms around to encourage action from the performers, and the orchestra returns the sound events.

The development of MIDI code and interfaces was ingenious in that it allowed electronic music to be produced with intuitive gestures, rather than laboriously typing code. The music industry was not interested in abstract symbols but in a gestural communication code that would help musicians play electronic instruments when performing onstage or in a studio. The concept of encoding gestures into electronics inputs/info is used in modern smartphones. We will talk more extensively about the importance of gestures in music in Chapter 15.

12.2 MIDI Networks: MIDI Devices, Ports, and Cables

MIDI devices communicate via electrical cables, simply called MIDI cables. MIDI messages are sent from an **Out-port** (output). They are then piped either to an **In-port** (input), where a machine uses the message, or to a **Thru-port**, which makes the message available to an **Out-port** of the same machine for further messaging to other machines.

12.3 Acoustics, Instruments, Music Software, and Creativity

A MIDI message consists of a 10-element sequence of zeros and ones which are called 'words'. These words are delimited by a zero startbit and a zero stopbit, and it takes 320 μsec (μsec=a microsecond, one millionth of a second) per word. MIDI messages are sequences of a few words with a fixed content anatomy. The first word is the status word, and the status character of the word is initiated by the first Bit: the *statusbit* one. The word encodes two information units: (1)

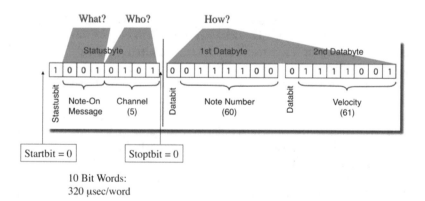

Fig. 12.2: The MIDI message anatomy.

what is being played, encoded in three Bits; and (2) who is playing, encoded in four Bits.

Let us first make the example shown in Figure 12.2. Here the first three Bits 001 encode the action of pressing down a key (Note-on, the note's onset action). The following four Bits 0101 tell us which "musician," called *channel*

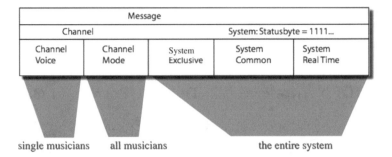

Fig. 12.3: The MIDI message types.

in the MIDI code, is playing the Note-on action; here it is channel 5, written in its binary representation 0101. With this information, the data words tell the musician which key to press down: The seven Bits in the first data word, namely 0111100, correspond to 60, the middle C on the piano. The second data word shows the seven-Bit binary number 1111001, corresponding to 61, and meaning "velocity." Here we see the gestural approach: "Velocity" means the velocity with which a keyboard player hits key number 60, and this is a parameter for loudness. The data words can specify any number in the interval $0 - 127$, so the velocity value is a middle loudness. The total time for such a message is three times 320 μsec, roughly one millisecond. This is not very short, and it can in fact happen that complex musical information flow can be heard in its serial structure.

As announced above, these numbers are still quite symbolic. The pitch associated with a key number is not given in the MIDI code, so it is up to the synthesizer to interpret a key number in terms of pitch. And velocity must also be specified by the synthesizer in order to become a loudness value. This is a typical situation of a connotative semiotic system: The MIDI Specification only decodes the Bit sequence, and then this result is taken as an expressive unit pointing to the acoustical contents via the synthesizer's signification engine.

The messages are divided into different categories (which are then encoded in the statusbyte). The most important are for musicians. These categories are shown in Figure 12.3.

Program Change, for example, describes the instrument that is going to be played. This is a typical situation where the meaning depends upon the synthesizer's settings. For the so-called General MIDI standard, these numbers have a fixed meaning, 0 being reserved for acoustic grand piano and 127 being assigned to gunshot.

The system messages are of three types: System Exclusive for messages that are reserved for specific information used by the industrial producer, like Yamaha, Casio, etc.; System common messages for general system information, such as the time position in an ongoing piece; and Real Time messages for time information (e.g. the timing clock that produces the ticks, see below about ticks).

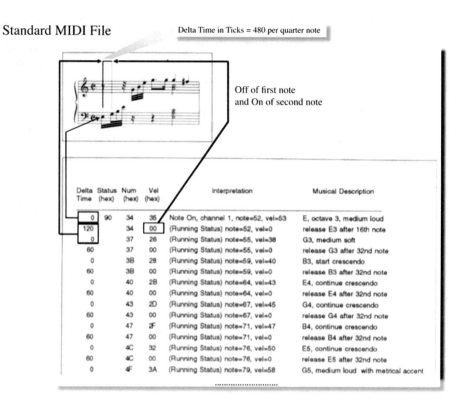

Fig. 12.4: A typical MIDI Standard File to see how this syntax looks in a concrete piece of music.

12.4 Time in MIDI, Standard MIDI Files

MIDI has a special treatment of time. It is not a physical time format, but counts time in multiples of *ticks*. The value of a tick is variable. Usually a tick means a 24th of a quarter note. But the physical duration of a tick must be defined on a special header. In the Standard MIDI Files, this header is called meta information and it tells us about the tempo that relates ticks to physical durations. Standard MIDI Files save MIDI information in order to play a recording back on a MIDI device. The *Standard MIDI File Formats* define the syntax of such files. Let us look at a typical Standard MIDI Files to see what this syntax looks like in a concrete piece of music (see Figure 12.4).

In the first column of the file's table, we see the delta time, which indicates in multiples of ticks the time delay between successive messages. The second column indicates the status byte in hexadecimal (base 16) numbers. The action is Note On, encoded by 1001, and the player is 0000 (first channel 0), yielding the hexadecimal pair 90. The next number is the hexadecimal key number 34, i.e., decimal key number $52 = 3 \times 16 + 4$—the E, first note in the score on top of the figure. Then we have the hexadecimal velocity value 53, i.e., the decimal velocity 53, medium loud. The next event is 120 ticks later and represents to play the end of the first note. Here we have a very economical convention, namely the "running status" command, which means not to change the status. This means that we again play that E but now with velocity 0! We play the note again but mute it. This is a creative use of the Note On message! When done with this note, we start the second note, G, without any time delay, i.e., the delta time is zero after the end of the first note, and so on.

13

Global Music

Summary. As a map of the world can be covered by a superposition of smaller maps of the different continents, music also has a global aspect. Each musical work can be seen as a whole (global) and its superposition of parts (local). The same can happen to tempo, where we can define hierarchies. Music is global in the sense that pieces and recordings can be shared through all the time and space in the world. However, a global level in the Renaissance or Baroque periods would have a much different meaning than today. If you asked Bach about globality, his answer would be totally different than from one of today's composers. Bach would be ecstatic if his piece reached a different continent. Today, many musicians celebrate being world stars via YouTube.

$$-\ \flat\ -$$

Music can influence technology because the musicians need tools to transform into reality their ideas. However, there is the inverse motion too: technology influences the way of making music. Let's now have a take at some tools of globalization.

13.1 The Synthesis Project on the Presto Software

Here, we illustrate the theory of global music on the jazz CD *Synthesis* which Mazzola recorded in 1990. Its entire structure—in harmony, rhythm, and melody—was deduced and constructed by use of the composition software Presto, see Figure 13.1 (written for Atari computers, but now also working on Atari emulations on Mac OS X). It was entirely based on 26 classes of three-element motives. Only the piano part was played by Mazzola, the entire bass and percussion part was played with synthesizers, driven by the Presto application via MIDI messages. During the production of this composition (*Synthesis* is a four-part, 45-minute piece), Mazzola never felt inhibited in his piano playing, on the contrary, it was a great pleasure to collaborate with complex

© Springer International Publishing AG 2016

G. Mazzola et al., *All About Music*, Computational Music Science,

DOI 10.1007/978-3-319-47334-5_13

structures of rhythm or melody. He was able to play with a complexity that human percussionists would never be able to play from a score. This composition was not recognized as a computer-generated music by the jazz critics. Most importantly, this enterprise was global in that it connected electronic music soft- and hardware to a real musician.

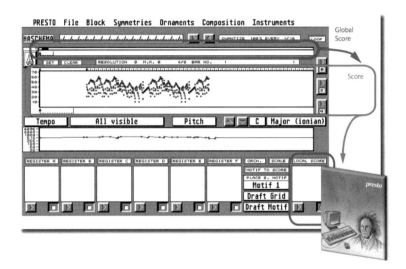

Fig. 13.1: The Presto user interface.

13.2 Time Hierarchies for Chopin's Impromptu op. 29

Music is something that exists within the parameters of space and time. More often than not, music constructs patterns. We can also say that patterns construct music. When we expand upon each musical gesture, we can see that each musical gesture is merely a rearrangement of movements in time. We make music when we make changes to these movements. We are also making music when we make adjustments to the sense of time. Let's talk about global time in music.

As we know from reading basic modern music notation, a note can take many faces. The noteheads and flags that we recognize are only fractions which are relative to a whole. Within Frederic Chopin's *Impromptu* op. 29, there are many different beats, and notes take on more fluid definitions (Figure 13.2). Time hierarchies are a subtle global structure in music. They are generated by taking "mother" times (such as the times set by a conductor) and generating "daughter" times for local variations. The execution of the grace notes may move differently in time than the half note (on beat 3 of the first measure

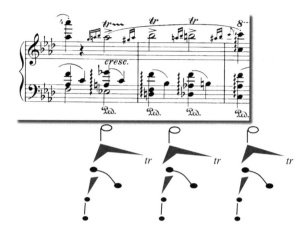

Fig. 13.2: A simple hierarchy breakdown within Chopin's *Impromptu for solo piano* op. 29.

shown.) Within the space-time of the half note, two quarter notes are played, creating a sense of time that lives within the larger beat. Music knows not only one time, but a whole genealogical tree of more and more local time levels.

13.3 Mystery Child

The composition "Mystery Child" from the album *Immaculate Concept* is an example of the Presto software completely altering the entire cultural context of a piece. With Presto, Mazzola transformed the Schumann piano piece *Bittendes Kind* from his *Kinderszenen* into something that sounds much more like Miles Davis' electric period than a romantic concerto. The change was produced by a deformation of the pitch and time positions of the original score, followed by instrumental changes. This "geometric" method can completely change the cultural character of a composition.

13.4 The Global Architecture of the Rubato Software

The Rubato Composer software[1] is a platfrom for musical analysis (harmonic, melodic, and rhythmical) and composition. It is implemented in Java programming environment. Java is platform invariant, which means that the same program can work on Mac, Windows, and Linux. Rubato's concept architecture is

---

[1] The older version of Rubato is implemented in Objective C and runs on Mac only. See www.rubato.org for more information on Rubato.

global since it enables the representation of most musical objects all over the world, from Western tradition to Japanese Noh music notation.

Rubato uses the definition of musical object types (denotators from mathematical music theory [41, 51]). Main concept structures are given by forms, a kind of general spaces, where denotators live (like points in such spaces). For example, if the form is pitch, a particular choice of value, for example A-4, is the base for the a pitch denotator.[2]

For the form loudness, the form type is string characters, and a specific denotator for loudness is given by a specific choice of characters, like *mf* to mean mezzoforte, as we can see in a musical score. A note denotator is defined as the list of denotators with coordinates for onset, pitch, duration, loudness and voice (selection of the timbre).

The formalism is derived from modern mathematics, namely from category theory. In a nutshell, we can say that a category is given by points and arrows, representing a set of objects and the morphisms between them. An entire category can be seen as an object, and the arrows between categories are the functors. This gives life to a nested system, which is a very powerful concept in computer programming. Category theory has been applied to music by Mazzola in his other works [41].

13.5 Braxton's Cosmic Compositions

> I know I'm an African-American, and I know I play the saxophone, but I'm not a jazz musician. I'm not a classical musician, either. My music is like my life: It's in between these areas. —Anthony Braxton

Anthony Braxton is a multi-talented musician who plays clarinet, saxophone, flute, and piano. Much of his music is based off of improvisation inspired by images. He composed scores which were just large, colorful paintings, such as in his *Falling River Music* project. These compositions were played by various small ensembles under Braxton's arrangement. One of his larger ensembles, the Tri-Centric Orchestra, gathered to record his opera *Trillium E* in 2010. All of his work is grounded in the belief that creativity must reject dichotomies. For example, Braxton supposes that music is not just composition and improvisation, it must also be intuition.

His earlier compositions featured smaller artistic diagrams that served as the title for each work. For example, his *Composition 76* was written in 1977 for three "multi-instrumentalists." The cover of the score features the visual title. The score itself is printed in full color and comes with original liner notes, composition notes, and a notation legend [6].

Braxton has written very global compositions for several planets. The first of these is his work titled *For Four Orchestras* (Figure 13.3). According to

---

[2] There is a correspondence between software structure in Java, with objects and classes, and the musical theory structure with denotators and forms.

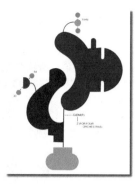

Fig. 13.3: Braxton's Composition 82 for four orchestras is meant to be played by 140 musicians.

Braxton himself, this work is the first completed work in a series of ten compositions that will involve the use of multiple-orchestralism and the dynamics of spacial activity. This work is scored for 160 musicians and has been designed to utilize both individual and collective sound-direction (in live performance). [7] Each orchestra is positioned near the corners of the performing space, and the audience is seated around and in between each orchestra. Each section of the performing space will give the listener a very different aspect of the music.

There are several different uses of time regulation in his compositions. Because of this, each area of the music has been designed to accent a particular principle. The time functions are:

1. Normal metric time (i.e. all orchestras functioning from the same pulse or time co-ordinate).
2. Multiple metric time (i.e. two or more orchestras in different tempos).
3. Elastic time (i.e. a time co-ordinate that is always accelerating or slowing).
4. Rubato time (i.e. a co-ordinate that allows the conductor to 'draw out' a given section of activity).

Braxton also mixes these time principles in different combinations. The problem of what this 'addition' poses for conducting is solved by the use of television monitors—connecting each conductor—so that sectional 'adjustments' can be regularly dealt with.

This work is supposed to sound multidimensional. The placement of activity in this project has been designed to totally utilize the spacial dynamics of the *quadraphonic technology* (four loudspeakers). It means that each orchestra will be heard coming from a separate speaker, and the mixture of events in a given section should give the sense of sound movement through space. This should also be apparent (though to a somewhat lesser extent) to stereo record players as well.

13.6 Machover's Brain Opera

The composer Tod Machover believes that anyone can make music. In 1996, he worked to make this true. With over fifty artists and scientists from the MIT Media Lab, he launched a project in 1996 called the *Brain Opera*. He sought to explore the mysterious interaction between music, language, and emotion. He developed a huge project which combined intuitive gestural applications in a place called the Mind Forest with live performers and the internet.

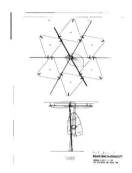

Fig. 13.4: A sketch of a tree from the Mind Forest.

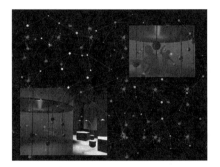

Fig. 13.5: An early artistic rendition of a rhythm tree.

The *Mind Forest* was a space that transformed gestures and vocal intonations into music and visuals. It created the illusion of walking into a musical brain space. The forest was constructed of 'Trees' (Figures 13.4, 13.5), Harmonic Driving units, Gesture Walls, and Melody Easels. Invisible sensors made the environment feel fluid and responsive, as opposed to forcibly futuristic. In the forest were fifteen "trees" into which people either sang or spoke. The tree would then transform the recording based upon how it analyzed and perceived the input, and use it in the *Brain Opera Performance.*

There was also a rhythm tree which was hand-sculpted from translucent rubber which connected more than 300 drum pads. Each would be able to recognize large variations in touch. Because they were all connected, each would be affected by its neighbors within the web to provide even more variation.

Using *Harmonic Driving* units (Figure 13.6), players could direct their way through a piece of music. As if in an arcade game, they would sit in a cubicle with a spring-mounted steering column before them. As the project notes describe, "the player's micro-steering, whether rhythmic and precise or sinuous and meandering, makes the music become sharp-edged or atmospheric."

Gesture Walls were created to allow full-body motion to control and perturb the music and imagery (Figures 13.7, 13.8). The player's feet would send a small electric signal through the floor plate which triggered the rest of sensors into activity.

Fig. 13.6: An early artistic rendition of a harmonic driving unit.

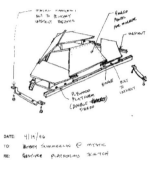

Fig. 13.7: An early sketch of a gesture wall.

Fig. 13.8: An early artistic rendition of a gesture wall.

The final feature of the Mind Forest was the *Melody Easels*. Much like finger-painting, users were able to drag their finger to create a melodic line. Their movement and touch create articulation and "timbral filigrees" that were represented in the sound and the visuals.

The Palette was the *Brain Opera* net instrument. It was a Java-based musical instrument which could be played from one's own home. On the dates of the *Brain Opera* performances, the website was activated such that all data would be sent to the Lincoln Center where the resulting MIDI would be played for the live audience (Figure 13.9). There were times in the performance of the Brain Opera that the internet performers were the primary focus on the stage.

The final performance was a three-movement, 45-minute composition. During the rendition, there were three performers who chose, transformed, and interpreted both composed music and sound improvised by the audience. The performers worked with the Gesture Wall, a Sensor Chair, and a Digital Baton. The Baton was able to sense direction and grip, the Chair converted movement into sound, and the Wall was a slight modification of the Wall from the *Mind Forest*. The musical frame for the first movement was Machover's

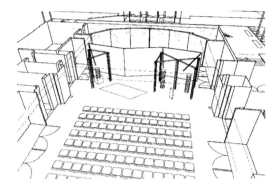

Fig. 13.9: A sketch of the Brain Opera performance space at the Lincoln Center

rendition of Bach's *Ricercare* mixed with the rhythm of spoken word and the audience's recordings. The second movement was heavily instrumental. At the beginning of the third movement, the "hyper-performers" on stage paused, and the internet performers took over. It developed into an alternating improvised harmonic flow. Finally, there was the "hyper-chorus climax" when every element was synthesized together in a final and grand recapitulation. The end product was a chaotic cloud of sound. However, the investigation itself of global music will forever be worthwhile.

13.7 The iPod and Tanaka's Malleable Mobile Music

More than ten years ago, having portable music devices may have seemed like a dream. With today's smartphones all around us, we can disconnect from the external surroundings and choose our own background music. Walking down the street, we can conduct our own soundtrack. Choosing to live in such an ever-present, artificial sound environment gives us a great sense of control. The effect of this is the epitome of our next topic. Music-making is egocentric. As artists, we seek to be the masters of collaborative musical creation. In 2004, Atau Tanaka initiated the concept of social musical collaboration all over the world, a vision that is now realistic with the omnipresence of smartphones.

Tanaka sought to extend music listening from a passive act to a participative activity. The system exploits ad-hoc wireless networks and mobility to allow a community of users to participate in the real-time creation of a single piece of music (Figure 13.10). Some detailed examples of Tanaka's plan were as follows:

> The evolution of the music comes from sub-conscious as well as volitional actions of the listeners. The intensity with which a listener holds the mobile device is translated into brightness of the music. The rhythm the user makes as he swings along with the music is captured and drives

the tempo through time-stretching techniques. The relative geographies of users in the group drives the mixing of the different musical modules. As a listening partner gets closer, their part is heard more prominently in the mix.

Fig. 13.10: A diagram for Tanaka's Malleable Mobile Music.

At that time, the technology and social acceptance just simply did not exist to make this a widespread ordeal. As of present times, *crowd-sourcing*[3] and *geolocation*[4] features are common, if not omnipresent, in cell phone use. If you go to a Starbucks, you may open Spotify on your smartphone to view the current playlist. Weather applications like Sunshine ask its users to report the current forecast to provide information to the rest of its audience. Facebook's Messenger and Apple's iMessage service both have integrated features which allow you to share your immediate location.

The explosive popularity of Apple's iPod throughout the early 2000s was the first realization of Tanaka's dream. In 2007, the iPod touch was unveiled which granted users the ability to connect to the internet via Wi-Fi. From here, the global aspect of music, creativity, and communication increased exponentially. Sing! Karaoke, an application produced by Smule, has turned music into a social networking platform. The app prompts users to sing along to popular songs and asks their friends to rate them. The music itself becomes a method of making contact between friends, ultimately making it a method of communication.

---

[3] Something that is crowd-sourced has asked participants to share information to supply content.

[4] Geolocation allows devices to be constantly aware of its location to help serve the user better.

13.8 Wolfram's Cellular Music Automata

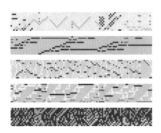

Fig. 13.11: Wolfram's cellular automata "scores"

In the 1980s, Stephen Wolfram studied ringtones which were made to resemble human-made music. His project asked the questions: Does creativity require humanness? Does it require our history and evolution? Or is creativity something which can rise from a system?

Wolfram's compositions stem from cellular automata algorithms. Cellular automata have grids of binary cells. These can be used to model a dynamic system. When one cell changes states, it affects the neighboring cells. He created these geometric tessellation structures to try to emulate creativity (Figure 13.11). Their sounds are as boring as machine-made stuff can be.

13.9 Mazzola's and Armangil's Transcultural Morphing Software

In a world full of remixes, it often seems that everything has already been created. Even language has been suffering from a lack of new possibilities. Modern Japanese borrows many commonplace words from English. For example, *kompyuu-ta* is just a cognate of *computer*. Music, like language, is only effective in the appropriate context. Imagine the possibilities if we could convert music to each listener's appropriate cultural context. It should be noted that culture is a deep, complex matter, and crossing between two cultures is even more challenging.

In Mazzola's and Alev Armangil's software, deformation algorithms are applied to deform cultural determinants (such as tonalities, rhythms, etc.) according to addressees in global communication. Although it appears to have changed on the surface, an inverse function would bring the piece back to where it began. For example, we can transform a Schumann piece to sound East Asian in tonality and rhythm, but it is still fundamentally based on Schumann. Then, we can use an inverse function to bring the piece back to its original state. This project was investigated by Mazzola and Alev Armangil in the context of Nokia's effort to globalize its market.

Part V

Embodiment

14

Recapitulation of the First Three Dimensions: Realities, Semiotics, Communication

Summary. Performers communicate content from various levels. In regards to music, this content is surmised of the score, its analysis, and its expressive meaning. The medium between a performer and their audience is through some sort of physical interface, such as a musical instrument. To produce the desired sounds, performers must execute precise gestures, using their training, knowledge of the score, and overall sensibility. This brings us to the peak of our discussion. All of the theoretical topics we've covered thus far can be summarily combined through gestures. There are many ways of making; thinking is a way of making. Listening is a very creative way of making, too. Gestures are another tool of creation. We say that they have a double ontology, existing inside the mental reality and also the physical reality. There is also a brain connection to the psychological reality. In this chapter we will explain how to contextualize gestures in the frame of realities, semiotics and communication.

14.1 Realities

As they pertain to music, the physical, psychological, and mental realities construct the playing field on which music progresses in all dimensions. Understanding them allows us to recognize boundaries and limitations in our musical processes.

The physical reality instantiates all possibilities, and consequently, all impossibilities. The fact that sound propagates through space over a period of time means that space and time are two fundamental facets of music. A sound wave cannot exist in an instant; only points can be recognized in the very smallest fraction of time. Sound is modeled by sinusoidal waves. To review this topic, as we will be discussing it modeled through human gestures, see Section 3.1.2.

The psychological reality throws a wildcard into the mix. It pushes and pulls at musical space with human perspective. Our ability to have, feel, and display emotions largely impacts the production of music. However, the psychological reality isn't just something which is felt, it can be analyzed and interpreted just like the others. Using electroencephalographs, we are able to study emotion empirically. Music's ability to impact our physiological state is part of what makes its communicative power so important.

The mental reality is an abstraction of observations of the physical reality. Through mathematics, we can see that there are symmetries within music. These patterns are important as we move into the geometry of gestural motion. Although we cannot objectively say what is beautiful, we can say that there is something beautiful to the mathematics behind music.

The combination of these realities present us with a confusing case. Artists exist at the convergence of all of these realities. Artists also work at the intersection of imaginary time and real time. As Paul Valéry suggested, musicians use continuous motion to interpret discrete symbols. Somehow, musicians transform black dots on a page into beautiful music which moves people to experience emotions. This is the ultimate mystery which sits at the very heart of creativity and gestures.

14.2 Semiotics

Understanding semiotics allows us to ascertain the origin of meaning within music. As much as music can be dissected by its sound anatomy and its psychological impacts, we can also observe it on a dimension which is entirely separate from the physical reality. In a way, understanding music through semiotics is much like grappling with linguistics. However, we must bear in mind that music is a separate entity than language. There may be similarities worth noting, but the differences are of utmost importance.

Hugo Riemann's harmony functions serve as another method of 'labeling' the function of musical elements. He proposed that every chord has some value which defines its role as something that is essentially tonic, subdominant, or dominant $\{T, S, D\}$. Ferdinand de Saussure supposed that there are six dichotomies by which we can examine musical attributes. Lastly, there is the Babushka principle which is the embedding of sign systems in one another which often develops in communication.

14.3 Communication

Communication requires at least two entities with a relationship which is supported on both ends. This is a dimension which is intrinsically bound to human interaction. One may argue that music is omnipresent in life. It is not just a tool we use to express ourselves, it may also be the poetical ego which constructs the framework from which musical content is received. The content then exists in a neutral level where it rests in between the two communicators. Frequently, communication utilizes the Babushka principle which allows for detailed analysis. Experiencing music likely involves several chains of communication.

We have discussed several examples of art, and their individual traits which are both artistic and not. We have reviewed the importance of music in film and theatre. Music provides the overall sense of motion to any plot, whether the story is extraordinary or stagnant. The presence of great music must also be contrasted with the lack of any music at all. While it may be a thought-provoking artistic choice, silence is a lifeless experience in the long run.

Infinite production is the future of music. With the expansion of virtual reality, three-dimensional sound immersion might be the new norm for musical enjoyment.

MIDI code is proof that music is headed toward infinite production. The discrete symbols which outline pitch and velocity may be applied to different voices, ultimately changing the entire mood and implication of a work. This is akin to changing the orchestration of a Mozart symphony to a 1940s big band. The realities of the same function (the score) would have changed entirely.

Global music is the expansion of infinite production into what could be termed instantaneous production. Many innovators like Tod Machover and Atau Tanaka sought to turn music from a passive listening activity into an active one. With failures at creativity, like Stephen Wolfram's cellular music automata, it is evident that human gestures are fundamental to meaningful creation.

15

The Need for a Gesture Theory in Music

Summary. Thinking is essentially a practice of making. We can also make in different ways, such as through actions which we call *gestures*. Western modern notation has its origin in Gregorian neumes, but music does not see such significant gestural advancements until more recent times. Through modernity, creators have come to rediscover gestures as fundamental components of artistic, and in particular musical, creation. We will analyze theories and thoughts, from Adorno to Hatten, passing through Chopin's performances and ending with Mazzola's contribution both as performer as well as theorist and research director.

15.1 Neumes and Musical Notation

Graphic representations of music, such as notes in the score, are simply frozen gestures. While some forms of notation do better jobs at allowing us to recognize the true nature of the gesture that supports such a symbol. It is difficult, however, to capture the nature of a four-dimensional motion (a change in spatial position over time) on a simple two-dimensional sheet of paper.

Starting with the Ancient Greeks, we saw *neumes*. These neumes (Figure 15.1) realized direction of pitch and inflection. Sometimes they would be placed directly above the words. Sometimes they were on an adjacent tablet. Lines were

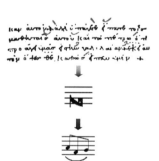

Fig. 15.1: The evolution of neumes to modern notation.

introduced to help give reference to the center pitch. Musicians had to translate the visual indication of space into the intervals between the notes. Eventually,

© Springer International Publishing AG 2016

G. Mazzola et al., *All About Music*, Computational Music Science,

DOI 10.1007/978-3-319-47334-5_15

a clef was added which allowed music to be written for different voices with greater ranges and stability of the center.

We already have music theory which helps us decipher what the scribblings on the page mean. We also have Schenkerian analysis to help us impose motion upon the static symbols on the page. What common music pedagogy lacks is a theory of gestures. We lack the understanding as to how and why we transform black dots on a page into melodies and harmonies that progress through time. If one wants to truly understand musicianship, one must—as much as one is encouraged to learn the literal meaning of the signs on the page—learn to express their implications.

15.2 Lewin, Adorno, and Hatten

Fig. 15.2: From left to right: David Lewin, Theodor W. Adorno, Robert S. Hatten

15.2.1 David Lewin

Music theorist, critic, and composer David Lewin (1933-2003) was born in New York City and studied the piano from a young age (Figure 15.2). His writings focused on mathematical group theory in music. He alluded to the gestural language in his book *Generalized Musical Intervals and Transformations* [32, p. 159]:

> If I am at s and wish to get to t, what characteristic gesture (e.g. member of STRANS) should I perform in order to arrive there?"... what sorts of admissible transformations in my space ... will do the best job?" ... This attitude is by and large the attitude of someone inside the music, as idealized dancer and/or singer. No external observer (analyst, listener) is needed.

This is a verbal analogy to gestures. We spell music using the same alphabet. When we look at the letters or notes on the page, we must ask ourselves *what gesture brings us from D to E?* On a clarinet, it's the lifting of the middle finger of your left hand. On a piano, you must pivot two fingers in order to dampen the sustain on the D to switch to striking the E key. On a marimba, you must strike the two separate wooden bars which vibrate on their own.

The notes on the page *are not music.* They are at most frozen gestures and at the least they are incomplete instructions which give musicians a starting point. Music must be experienced, but no experience is possible from ink blots on a page. Imagine someone showing you a Rorschach test and calling it the score. Only gestures create the experience. They are only realized if someone (or perhaps, something) does it. That is why music is a *system* of signs. There is content, but in order for it to become an expression of sorts, it must first undergo signification.

In *Musical Form and Transformations* [33, p. 41]: Lewin continues in this language[1]:

> The relations underlying example 2.5 exist outside human time in an abstract universe of quasi-spatial potentialities; they cannot be manifested in music except through human gestures that move through chronological time.

Here, Lewin continues to argue that music must be realized for it to exist. The simple intention of making music has little value compared to the actual gestural motions of making music. Lewin preaches the gestural language, although his theories work with classical set-theoretical functions.

15.2.2 Theodor W. Adorno

Composer, philosopher, and sociologist Theodor W. Adorno (1903-1969) was born to a family of singers in 1903 (Figure 15.2). He was revered as a young prodigy at piano. Adorno died in 1969, leaving a vast influence across music, especially that of Karlheinz Stockhausen. Adorno held a very specific view of what serious music should be. He insisted that serious music must cause the listener to contemplate. Listeners should not be able to immediately predict what happens next. He believed that music should stimulate the mind. Essentially, he would not be fond of Top 40 radio. Music was not inherently supposed to bring people together; it was supposed to induce introspection. In *Towards a Theory of Musical Reproduction* [1] Adorno wrote:

> Correspondingly the task of the interpreter would be to consider the notes until they are transformed into original manuscripts under the insistent eye of the observer; however not as images of the author's

---

[1] Using a pitch class set network for Stockhausen's *Klavierstück III*.

emotion, they are also such, but only accidentally, but as the seismographic curves, which the body has left to the music in its gestural vibrations.

In other words, music should be so well performed according to the score, that one should be able to transcribe the gestures. Theoretically, an "accurate" and precise performance would lead to one being able to literally rewrite the score, a skill which Mozart was rumored to have. However, Adorno argues something further. It is unlikely that an audience member would be able to describe the exact thoughts and emotional intentions of the composer. Therefore, it is most essential that an audience member can understand the curves and gestures that the composer intended. It is then the performer's duty to serve as a bridge. To Adorno, music, even serious, contemplative music, was meant to be gestural in nature.

15.2.3 Robert S. Hatten

Robert S. Hatten is a music theorist and professor working at many institutions (Figure 15.2). In his book *Interpreting Musical Gestures, Topics, and Tropes* [23] he defines gestures and applies this conceptualization to analyzing the interpretations of classical Viennese compositions. He comes to a remarkable conclusion:

> Given the importance of gesture to interpretation, why do we not have a comprehensive theory of gesture in music?

His argument stands strongly on its own. If pedagogy insists that Schenkerian analysis is important to music, then we must assume that physical motion is as important to music as implied motion from tonic to dominant to tonic $I \rightarrow V \rightarrow I$. The only way to give kinetic life to music is through gestures. Just as we study the meaning of black dots in lines within the meaning of music, the study of gestures which bring the signs into artistic communication must be at least equally important.

15.3 Lang Lang and Marquese 'Nonstop' Scott Performing Chopin's *Etude* no. 12

In 2012, world renowned pianist Lang Lang and dancer Marquese 'Nonstop' Scott collaborated to film a gestural music video. Lang Lang performed Chopin's *Etude* no. 12 as Scott interpreted through dance. Through the video, it is obvious that having two performers interpret the same material led to a deeper understanding than a soloist alone. As Lewin, Adorno, and Hatten might say, the gestures brought us closer to the neutral content, the score.

On the other hand, one might argue that Scott was simply dancing to the dramatic sounds he heard from Lang's performance. This would imply that

Scott's performance is a step away in the communication system from Chopin's neutral content. However, it is still fundamentally an interpretation of this etude. Secondary sources of information may provide just as much insight as the primary source. In fact, it is possible that the secondary source, by its very nature of being an interpretation, could have an even more apparent meaning than the primary source.

Scott's contribution to the video is vital to the artistic value of the work, for there seems to be something missing in Lang's performance. Lang is very mechanical and only produces gestures through his hands. While this is still valid gestural motion, it does not have the impact of Scott's full-body contortions. A performer must add content through their interpretation to make it more interesting than the score alone.

15.4 Mazzola's Contributions

Guerino Mazzola is a long-established writer within the mathematical music theory world. His works include texts and multimedia which educate consumers on the importance of a comprehensive and gestural view of music and arts.

15.4.1 Teak Leaves at the Temples Movie

Teak Leaves at the Temples was filmed among a variety of communities in Indonesia. The documentary presents a free jazz collaboration by Western jazz artists Guerino Mazzola on piano, Heinz Geisser on percussion, and Norris Jones (art name: Sirone) on double bass, along with the community of Lima Gunung and the Sono Seni Ensemble, and the art community of the Boko, Prambanan and Borobudur temples. The scenes in the film are mostly taken around these three ancient temples. Drawing reference to the "jazz" of everyday life, director Nugroho juxtaposes the domestic life of the village communities with the free form musical structure. In essence, free jazz parallels the lives of these communities: They adjust to the natural forces which affect their surroundings and are flexible to improvise when the need arises. Teak Leaves at the Temple is a fascinating piece of tapestry, weaving together the diverse cultural influences of Indonesia. For the authors, an important message of this film is that, like jazz, we are free to shape our lives; we do not follow pre-determined forms.

The 'Western' musical component involved the trio Geisser, Mazzola and Jones. The main message of their collaboration with Indonesian musicians is that communication and understanding can help to overcome differences and start collaborations.

A point of convergence between cultures in this context was musical communication through gestures. Understanding on the gestural level transcends boundaries of language, appealing to a shared nucleus of communication in the frame of musical improvisation that can be found in music. The two groups could communicate on a purely gestural level, without needing explanation or direction.

The two groups (Mazzola and the Indonesian group) dress differently and speak different languages. Though the traditional performance attire between the two groups was different, they were able to transcend these cultural differences and communicate on the gestural level.

Several sequences comprising comparisons and conceptual superpositions involved the rhythm of people making sculptures, followed by the noise of the flowing water. There is a music inside 'musical sessions,' but music is also present as a part of the human body and natural environment. For example, Ismanto (one of the Indonesian artists living at the Borobudur temple) notes that "music is... working, sculpting, hoeing, plowing and dancing. Dancing is also working. So is music." Organizing the musical realities of these different performances is also a form of musical creation, very frequent in soundtracks. The idea is to merge these musical realities, bringing together the voices of external reference (nature and sculpture) and strict musical form.

In order to do so, some shared gestural paradigms inside artistic improvisation are required. Collaboration enriches traditional music. For example, the jazz movement opened a new perspective to the local musicians. For those living on a mountain, this presented a unique opportunity to play music with international musicians from Switzerland. This collaboration inspired them to produce creative works based on the free improvisation of jazz music. Improvisation as a continuous creation of new music can be compared to the visual art tradition of the mandalas (Figure 15.3), where different images are quickly destroyed and made again, in an infinite process.[2] In the words of Mazzola: we are 'pointing to the infiniteness of the universe.' It all starts with the intrinsic meaning of the events that happen in our lives. People can draw from this infiniteness in their own way, producing beautiful forms, full of the body of time.

In the movie, one after the other, the images of a resting dancer and then statues of stone were shown. And then, the images are replaced by an announcer describing the destruction caused by a earthquake. The scene ends with the temples, again with the stone. The television in the movie didn't show the image of the earthquake happening, but only its final effect. Like the gestures that 'happened,' leaving the 'static' result, we are faced with the conundrum of creating art in the frame of catastrophes.

We can also witness the phenomenon of art after catastrophe in the West. After the bombing that happened during the war, the conductor Arturo Toscanini wanted to reopen the theater La Scala in Milan, and resume conducting even though many buildings were still in ruins. This example shows that art can be a reaction to negative things and in turn a source of motivation

---

[2] A mandala is an intricate design of a spiritual symbol in Buddhism that represents the universe. However, it has become a generic term for symbols used to represent the cosmos and the universe.

Fig. 15.3: The vibrant figure of the mandala.

for people. We can cite, in this context, the words written in the pediment of the opera theater Teatro Massimo of Palermo, showing a similar sentiment in Italian culture: "L'arte rinnova i popoli e ne rivela la vita. Vano delle scene il diletto ove non miri a preparar l'avvenire" (The art renews people and reveals their life. The art is useless if it is not used to prepare for the future [of people]). We can envisage here the gesture of recreating a destroyed environment, piece after piece, and in this situation, art has an important motivational role for people. A very significant scene in the movie shows a little child, who—instead of staying without moving and crying—helps build the artwork by contributing a little decorated piece of rubble.

Another theme in Western art is the idea of art as eternal. Ugo Foscolo expresses in *De' Sepolcri*, 'Onde d'Elettra tua resti la fama' (In order to keep the memory of your Electra), and Horatius says 'Exegi monumentum aere perennius' (I've built an [artwork] more eternal of the air). He emphasizes the eternity of the artwork and its importance for the spiritual nature of humans. In the movie, after a sequence of images of destructions, came some music from the ruins. About the reason behind art, in the movie, Ismanto said:

> Why do we perform? It all starts from how we perceive things. Whatever happens around us has its own meaning and everybody has their own way of seeing things. People can be drawn into sorrow by their own minds. But when we perceive an earthquake, a mountain eruption or a mud flood, we know that there are good things from it. It will inspire a work of art like my Buddha in the Arupadhatu position, which was exhibited during the art fair. It is all about perfection that is inspired by those disasters.

We can say that it is also a way to react to tragedy by using art to have the courage for struggling and staying alive.

In a scene of the movie, Mazzola was listening to and watching the other musicians, in particular the singer. He was following the musical pattern by moving his hands in the air. Then, he played real music at the piano, with the same flowing gestures. It gives an idea of the progressive generation of artwork.

In another sequence, Mazzola talked about the development of art, taking ideas from the *Borobudur* temple, in particular starting from the observation of human scenes depicted in the bas-reliefs, see Figure 15.4. The represented

Fig. 15.4: A bas-relief from the the *Borobudur* temple.

human gestures are 'frozen' inside the sculpture. He talks about time: "Time begins as very abstract, in order for life and time to interact they must take a form (the form of gestures). It's a big musical score of human life," he says. The realization of artwork as translation of thought inside a 'material' shape is again referenced by Ismanto.

> It's only a question of how we talk to reality, it is not reality itself that matters. Using the words of Mazzola, "It is the imaginary reality that is important to artistic creation." Time would be quite an abstract line without life. How can we create form? We have to perform gestures by the movement of human bodies. It is the human gestures which shape the body and time.

In fact, giving a shape to a stone means to transfer the mental image to a physical object (action of the artist). In this way there is a reference to the transition from the imaginary time of mental reality, to the real time of physical reality. See Chapter 16 about gestures for more detail about these concepts. In further reference to sculpture and art making, Ismanto describes the relationship between artist's imaginary reality and the physical reality of nature:

> I strongly hit the stones, they stay... I touch them gently, they give me a reply, they can tell stories. They can answer all my questions. Until one day they started influencing my life. I wanted to live like a stone. I am so influenced by their presence. They can also be like an altar, as a

symbol, when we face God. I often ask the stones because I can get an answer from them... Music is working, sculpting, hoeing, plowing and dancing. Dancing is also working. So is music. It is really fun to see Mazzola playing the piano. Free, just flows. That is what he does every day. What a life!

We could say that, in some way, nature answers all the questions. In one scene free jazz music was represented by the flowing river water. Ismanto talked about 'flow' in Mazzola's piano playing while listening to the sound of the flowing water of a river. In this frame of flow, we can think about the equilibrium between man and nature, seeing nature as a source of artistic inspiration.

Dealing more generally with culture, we discover the relationship between symbols of religion and musical-style. These two concepts have been intertwined across all cultures. There are several examples of specific musical and religious symbols that appear with the same meaning across different cultures. For example, the musical gesture of an ascending melody points to a higher power.

In the case of voice, it is possible to uncover analogies between the Western and Indonesian techniques. For example, in one performance the singer started by jumping between two notes in an Indonesian melodic interval, then successively decreasing the distance between these notes until they converged, ending up with little pitch variations corresponding to the Western vibrato. This process can be seen as an example of 'cultural modulation' with music.

The movie begins and ends with the image of leaves, mixed with the slow movement of the dancer, contrasting with the fast tempo of the music—gestures in 'slow-motion.' The final gesture is dedicated to nature: pointing to the moon, and letting everything go to the new creation.

15.4.2 Books About Gesture Theory in Music

1. Guerino Mazzola: La vérité du beau dans la musique [43]
   The French book is the first to present the mathematical theory of gestures to a broad public. The book is written in everyday language and deals with the philosophy of mathematical theories of music in harmony, counterpoint, performance, and gesture.
2. Guerino Mazzola and Paul Cherlin: Flow, Gesture, and Spaces in Free Jazz [45]
   This book deals with free jazz as a collaborative art in open spaces, where gestural communication can build a flowing energy.
3. Guerino Mazzola: Musical Performance [46]
   This book is the first comprehensive presentation of modern performance theory, including gestural aspects thereof.
4. Guerino Mazzola, Joomi Park, Florian Thalmann: Musical Creativity [47]
   Musical creativity as a theory is presented in this book, and gestural rationales and inspiration for creativity are discussed and exemplified, among others for Beethoven's Sonata op. 109.

5. Guerino Mazzola et al: The Topos of Music III: Gesture [42]
   This book is a thorough investigation of mathematical gesture theory. It complements the now classical *Topos of Music* of the same author.
6. Matthew Rahaim: Musicking Bodies [59]
   This book is a groundbreaking contribution to the study of Indian music, musical performance, and music cognition, with a strong emphasis on gestural expressivity.
7. Mark Leman and Rolf Inge Godøy (eds): Musical Gestures [31]
   This book is a collection of essays addressing the fundamental issues of gesture research in relation to music.
8. Alexandra Pierce: Deepening Musical Performance Through Movement [58]
   Pierce's book is about theory and practice of embodied musical interpretation.
9. Robert Hatten: Interpreting Musical Gestures, Topics, and Tropes [23]
   In addition to expanding theories of markedness, topics, and tropes, Hatten offers a fresh contribution to the understanding of musical gestures, as grounded in biological, psychological, cultural, and music-stylistic competencies. By focusing on gestures, topics, tropes, and their interaction in the music of Mozart, Beethoven, and Schubert, Hatten demonstrates the power and elegance of synthetic structures and emergent meanings within a changing Viennese Classical style.

16

Frege's Prison of Functions

Summary. Masterful piano players captivate audiences not only with sound, but with the spectacle of their movements, which also influence the quality of the sound. Many interesting gestures involve rotations in space. We can use a formula to describe a rotation, but the formula denies us the understanding of its essence. Mathematical abstraction neglects the gestural aspect; it takes away from the preceding gestural essence, hiding its nature of movement that connects points in spaces. The whole movement is hidden at the very least, or destroyed at the worst. We should be able to understand a formula via its unfolding in gestures, as in the case of rotation. Several important mathematical topics can be re-thought under the light of gestures. Complex numbers can also be seen as the result of the gestural action of rotation. We will use the imagery of mirrors and vampires to help us understand. We end the chapter with the distinction between imaginary time and real time, the two components of the complex time.

$$-\,\&\,-$$

Let's imagine an arrow bringing points of a set A to points of a set B, a very intuitive representation of a *function*. How can objects of A can be 'transformed' into objects of B? Via some magic formula? We will discuss formulas, not magic (except our later reference to mathematical vampires). Does a function involve a teleportation of objects between sets, or a slow and continuous transformation of the ones into the others?

We can ask, in this context of A to B, what is the difference between 'teleportation' and 'transformation.' In the first case, we just see two different states, an initial and a final one of a system in two different places. In the second case, we take into account the continuity, the (perhaps infinite) intermediate passages that transform an object of A in an object of B. For an intuitive explanation of this concept, see Figure 16.1. We can also think of *chronophotography*, a sequence of successive photo shots, showing the movement of the object, as partially superimposed stills of a movie. Chonophotography gives us

© Springer International Publishing AG 2016
G. Mazzola et al., *All About Music*, Computational Music Science,
DOI 10.1007/978-3-319-47334-5_16

an intuitive concept of gesture as objectivized trajectory.[1] In Gottlob Frege's "functional" formalism we are not interested in *what happens* between A and B, but just to two given points, an input and an output, like *a beginning and an end, without a story in between.*

Fig. 16.1: Giacomo Balla's *The hand of the violinist* well illustrates the concept of continuous transformation from a starting point to an ending point, hidden in the definition of function. This concept is also at the basis of the definition of gesture as a system of continuous curves connecting points in space and time.

Despite this problem, thinking with arrows can be a powerful tool, as shown by contemporary mathematics and philosophy, based on the formalism of category theory (see Sectionrubatosoftware), but it should be used carefully— knowing what it is hiding. In fact, we must take care of the hidden gestures, actions that are underlined by such a formalism. When we use arrows, we should being conscious that, in formulas, gestures are made 'compact.' As a performer reading the symbolic expressions in a score and transforming the hidden instructions of movement into real, physical gesture, that finally leads to a sound result, understanding formulas means also to *unfold* their hidden gestural content. A musical score does not directly show the music, but it must be interpreted by a trained musician in order to be transformed (and its content transferred) into sounds. Like a score that can be seen as a container of 'frozen' gesture, we can also see math as a way to 'freeze' gestures into the symbolic reality of formulas. These formulas must be 'unfrozen' or 'unfolded' to reveal the nature of their content. For this reason, the mathematician and philosopher Gilles Châtelet [11] criticized a 'teleportation' view of functions that forgets their gestural content: functions are 'phantoms' of the gesture. See Reference [48] for a thorough discussion of gestures in music and mathematics.

---

[1] In mathematical gesture theory, the interest is shifted from the object moving to its trajectory.

16.1 Matrix Encapsulation of Geometric Rotation

A very important case of a hidden movement inside the symbolic reality of formulas is the rotation. If we have a column vector (in a nutshell: a vertical set of numbers representing the value for each coordinate—position of an object in a space), and we applying to it a particular matrix (a set of numbers arranged in a rectangle of rows and columns), we can get another column vector, corresponding to the new position of the object in the space. The *rotation* matrix, for example, allows such a change of position. However, as explained in the previous section, the rotation is not already 'visible' from the formula: this last one must be interpreted[2] in order to find the 'moving instruction' of a rotation. In this sense we can also talk about 'encapsulation' of the geometric rotation inside a matrix.

16.2 Dracula and the Imaginary Numbers: How to Solve the Singularity of Real Number Negation

Perhaps you're aware that vampires are absolutely disgusted by garlic. Perhaps you're also aware that a vampire's reflection cannot be seen in a mirror. What relevance does such an elusive entity have to the study of music? We propose that the existence of vampires is a parallel to music.

The best way to imagine this is to picture space like a book, where the leafing of pages represents rotation. In fact, we can use the gestural image of rotation to explain the necessity for complex numbers. A complex number is given by two parts, a real part, and an imaginary one. The number in the imaginary part is preceded by 'i,' the imaginary unit, equal to the square root of minus one. Complex numbers are represented, not in one line as the real numbers are, but in a plane, with the horizontal axis corresponding to the line for real numbers (real part), and a vertical axis for the imaginary part.

Let us think of the case of a positive number, for example $+3$, that is sent into -3. If we make a transformation only on the real line, the number will 'disappear'in 0, see Figure 16.2. If we make a rotation with respect to the vertical axis, there is no singularity in 0, it means that we always have a 'mirrored image' different from zero. We would have a situation like the vampires, who cannot see themselves in the mirror. In fact, their real part is equal to zero, and so, vampires are purely imaginary entities.

The passage from a positive real number x to the negative $-x$ cannot be thought as a continuous movement, because each continuous curve would pass through the origin $(x = 0)$, becoming singular. The mirroring operation, if seen only from the perspective of real numbers, is a teleportation in the Fregean sense. However, if we add the vertical axis of imaginary numbers (Im),

---

[2] This is a tedious work, related to exhibit eigenvalues and corresponding eigenvectors (rotational axes) for the given matrix.

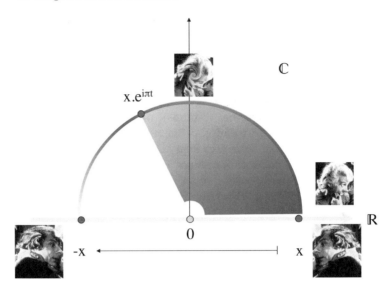

Fig. 16.2: Rotation in the complex plane. While real humans (here Roman Polanski in the movie *The Fearless Vampire Killers*) can be reflected also via a rotation in the complex plane, vampires get stuck in the imaginary axis, you cannot see them in a mirror.

obtaining the plane of complex numbers, the mirroring can simply be explained via a rotation with respect to the Im axis. In $i = \sqrt{-1}$ the real part is zero, and in this point, the origin, corresponding to half-path of the mirroring (the mirror!), only the imaginary component is non-null.

To conclude, we can argue that, despite the fact that garlic is real and Dracula is imaginary, mixing garlic and vampires is something very... complex.

16.3 Imaginary Time

We will see now how it is possible to extend the formalism of physics to include the reality of the *mind*. And, to do so, we need the complex number-concept just defined.

The idea to extend the formalism of physics to the mind has been proposed by the theoretical physicist Roger Penrose. [56] Another theoretical physicist, Stephen Hawking, used the complex time to solve singularities in the Big Bang. [24]

Creatively combining the two ideas, it has been proposed to use complex time to extend physics to include the mind, opening the walls of the restriction of formalism. [36]

In few words, referring to music, we can see (complex) time as being constituted by two components, the real and imaginary one, where the real time flows in the real musical performance, and the imaginary time is the time of the mind, represented in the symbolic time of a score, with its *tempo* indication, its metronome mark, and its duration figures of notes. In general, the transformation of imaginary into real time during a musical performance is not linear: you can think of a performer who, while rehearsing, goes back to a previous section, or changes the tempo, not being satisfied by the sound effect, for example. However, to build a first model of each step of musical performance, from the score to the final concert, such a simplification is necessary.[3]

The concept of complex time seems to be powerful also beyond music. We can think for example of the concept of potentiality and actuality in Aristotele's philosophy, with the difference, in our musical case, that both symbolic and physical reality of score and performance have *their own identity as realities*, and the first is not only a 'preparation' of the second. We can also use complex time to solve the Cartesian duality between *res cogitans* and *res extensa*, the threedimensional physical space being seen as the shared part of the two ontologies.

16.4 Other Hidden Concepts in Math

16.4.1 The Density of Real Numbers

There are an infinity of numbers. We can never hope to count to infinity, nor will any machine ever reach it. There is no cap. There is no end. This is a fact that you've likely toyed with throughout any calculus you may have studied. We can hardly begin to understand every implication of infinity ourselves, much less explain it to you, our readers. The fact that the ceiling can never be reached is all that we need to bear in mind.

Now, if I were to ask you to count to ten right now, you probably thought *one, two, three, four...* all the way to *ten.* You thought of ten distinct entities called natural numbers. But how many real numbers exist between one and ten? An infinity. That's because, technically speaking, you had to reach 1.5 before you reached 2, 1.25 before 1.5, 1.125 before 1.25 and so on. When do you actually ever reach two?

Looking at the face of an analog clock, you see the minute hand slowly crawling from the one, which indicates five minutes past the hour, to the two, which indicates ten minutes past the hour. Well, no one enjoys staring at the wall, so you've lost track of time momentarily. You look back up at the wall at the minute hand has already reached the three, which indicates a quarter past the hour. Surely, the minute hand must have passed the two, but how could it have if there is an infinity of numbers between one and two? First it had to reach 1.9, and then 1.99, and then 1.999, and then 1.9999. At some point, the

---

[3] *Imaginary Time* is also the title of a recent DVD by the trio Guerino Mazzola, Heinz Geisser and Shiro Onuma, discussed in Section 20.2.

difference in time becomes nearly imperceptible to humans, but the conundrum still stands, nonetheless.

We measure time numerically in seconds. Seconds are divisible. Therefore, we can distinguish between the winner of a race by milliseconds. However, time is not fixed. According to Albert Einstein's Special Theory of Relativity, it is varies between different inertial systems, therefore it ultimately defined by its perceivers. Here, we introduce our two musicians, a drummer and a bassist. The backbone of any modern small ensemble, spanning jazz to rock to disco pop, these two individuals constitute what is commonly referred to as the rhythm section. Their most important contribution to the rest of the ensemble is to keep a steady pulse or a 'sense of time.' Somehow, they must execute certain beats precisely in sync to create the flow of the song. However, the way that they go about this is anything but precise, musical time is elastic.

Making music is a game of anticipation. Our instrumentalists may use visual and acoustic indicators to communicate, such as bobbing a head or tapping a foot, but these actions are limited by the speed of light, sound, and the human neural networks. A conductor can stand on stage and wave his arms in an effort to show his flow of time through the sounds of a symphony. However, his gestures will only ever be interpreted after they have occurred.

Music is a convergence of the infinity of numbers and their finite, quantifying nature. A gradual *ritardando* on the last beat of a measure could extend for as long as the performers decided, but eventually they play the next note, and so the progression through time marches on.

You might say that in music, you are momentarily creating a new frame of reference for time.

16.4.2 Zeno's Paradox

Greek philosopher Zeno of Elea described a paradoxon which is a great example of the problem of traveling through the infinity of every time moment. He proposed that the hero Achilles was racing a tortoise. Because Achilles assumed that the tortoise would undoubtedly lose in a standard footrace, he granted it a head start of 100 meters. Once the tortoise has reached 100 meters, Achilles dashes forward. In the time that Achilles reaches 100 meters, the tortoise has crawled forward another 10 meters. We would think that Achilles would eventually pass the tortoise. However, if we imagine that for every point the tortoise moves across, Achilles must also reach that point, then Achilles will never actually reach the tortoise, as there are an infinity of points. Because there are infinite locations where Achilles must pass until he reaches the tortoise, he will take infinitely long (essentially never) to reach the tortoise.

The implications of this are tremendous. We try to imagine real time, but it simply cannot be done without making a leap of faith. We must assume that at some point we reach the next place in line to keep moving forward.

Of course, this paradoxon has it solution, namely in the obervation that the infinite number of intermediate steps in this race does not mean the time to cross all of them is also infinite. Let us make a simple example of this situation: Take the first number one, then add $1/2$, then add $1/4$, then $1/8$, etc. add all powers of $1/2$. This is an infinite sequence of steps, but the total of the summation is exactly 2: You go from on to 1.5, then to 1.75 etc, you always add half of the remainder to 2. Zenon's conclusion was the wrong implication that an infinite number of logical steps implies an infinite *physical* time that is defined by these steps. The logical gesture in *symbolic time* and the corresponding time gesture in *real time* are very different things.

17

Music Without Scores

Summary.

A score is to a symphony as a formula is to a rotation; it has all of the instructions, but it does not contain any gestures within itself. If the formula is essential to a rotation, then one may also conclude that the score is critical to a symphony. This leads us to ask, can music exist without the score? The answer is yes. Given that real music is derived from gestures, then we must study the gestures that inherently exist. If we can understand gestures without a score, we can only better understand gestures that are accompanied by a score. However, in the score's existence, there is no action or progression of time. Such is the belief of many musicians trained both classically and in jazz. In this chapter, we will examine the works of free-jazz pianist Cecil Taylor and the works of robots.

Fig. 17.1: Cecil Taylor: *Burning Poles.*

$$-\; \&\; -$$

17.1 Cecil Taylor: Burning Poles

Although classical Western music relies mostly on scores, there is a large culture of music which does not. This is especially the case for improvised music. The free-jazz pianist Cecil Taylor creates his music from his gestural dance. Taylor has said that he tries to imitate a dancer's leaps and movements with his hands. He performs a dance of his hands, but also of his entire body. In the video *Burning Poles,* he performs an entire work while dancing around the piano (Figure 17.1). He deduces everything into a gestural dance, without the aid of lead sheets. If you were to attempt to traditionally transcribe Cecil Taylor's performance, you would be left with an extensive section of "silence." In the video, he dances around the piano while vocalizing and gesturalizing. To Taylor,

© Springer International Publishing AG 2016

G. Mazzola et al., *All About Music*, Computational Music Science,

DOI 10.1007/978-3-319-47334-5_17

the piano is as much a part of the performance as his own body. Both must be prepared and approached carefully. In a provocative way, Taylor stated that

In Western music, the body plays no role.

He is very deliberate in this harsh statement. It is a criticism of the current state of music in the Western culture. There is next to no emphasis on gestures and the use of one's body in the popular music scenes. Electronic music can be produced by producer at a computer workstation. Today's most renowned orchestras are typically comprised of 100 strong musicians, but the spectacle is in the grandiosity of the sound, not the gestures.[1] With the prevalence of digital music, from streaming to CDs to MP3s, the little gestural enjoyment of music that might exist is almost entirely erased, limiting musical expression to that of sonic expression only.

Cecil Taylor's music must live beyond and without a score, because the *gestures* are his content. An accurate transcription of *Burning Poles* would involve a description of his gestures, either verbal or visual. His message is important, not for his creativity alone, but because he places such great importance on gestures.

17.2 Artificial Embodied Intelligence: Cheap Design

Fig. 17.2: Rolf Pfeifer's stumpy robot in "cheap design".

Embodied Artificial Intelligence comes from the insight that the brain and the neural networks alone cannot fully explain intelligence. However, this new direction is ultimately farther from intelligent behavior. This new A.I. research tries to reconstruct intelligent behavior from bodily movement alone (Figure 17.2). This perspective is also called cheap design, because its proponents believe that intelligent behavior can be a consequence of simple mechanical arrangements of limbs. This is, however, an oversimplification of cognition. The body alone does not guarantee any intelligent behavior. What is needed is a connection between the body's movements and the cognitive activity. This connection is probably achieved through gestures, which connect the body with cognition. Recall that Jean Cavaillès defined understanding as "catching the gesture and being able to continue." [10]

So how does visual content get translated into physical movement? This is a very fluid, innate concept for humans. However, it is not so simple for mechanisms and computers. For humans, the time needed to recognize a spatial rotation, is the time needed to make the gestural rotation. [12] Intelligent

---

[1] Conservatory training is partially to blame for this phenomenon. Classical musicians are frequently trained to play with "predictable" drama and to *not* be creative in their bodily motions.

behavior requires the use of imagination. At the time of writing, Google's Deep-Dream is the closest thing to a computerized imagination. Intelligent behavior is not just the mechanical movement of limbs. There is no such thing as 'muscle memory'; this completely discredits the sheer power of the nervous system. The memory truly lies with the neurons, which learn the gestures. This will be discussed more extensively in Section 18.2.

17.3 The Hand Computer Graphics as a Gestural Challenge

As an experiment with cheap embodiment, we now study an actual simulation, the movement of a pianist's hands. Referring back to the homunculus and the amount of mental space dedicated to the hands, we can conclude that hand gestures are complicated. (Refer to section 18.2) Music can shape more than just the area of musical perception in the brain. It can also modify something as general as hearing and physical movement.

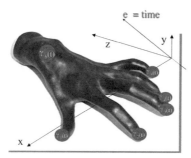

Fig. 17.3: A diagram of the points in relation to a hand.

In a computer graphics software, six gestural curves were modeled in space-time $(x, y, z; e)$, see Figure 17.3. There were five curves for each fingertip and one for the movement of the wrist and body of the hand. The hand was given a staff of music to interpret, the fingerings were calculated, and the simulation was rendered. The computer was the one who decided *how* each finger will be placed at each certain point in space and time. It learned through calculation and executed gesturally. This is one step closer to, but still one step away from, natural human learning processes. We hope that robots may one day be able to learn gesturally as well.

17.4 Robotics for (Musical) Gestures: Asimo & Co.

ASIMO, "The World's Most Advanced Humanoid Robot", was pioneered by Honda (Figure 17.4). Its name is derived from its longer title, an *A*dvanced *S*tep in *I*nnovative *MO*bility. The 2011 model is four feet and three inches (130 cm) tall, with similar proportions to a human child. Built from plastic and a magnesium alloy, ASIMO weighs in at about 105 pounds (48 kg). The backpack which contains the lithium ion battery weighs about 13 pounds (6 kg). It has 13 degrees of freedom per hand, with a total of 57 encompassing its entire body.

Fig. 17.4: ASIMO.

ASIMO was created to help people with limited mobility. However, his actions speak volumes of insight. Creating robots like this are a huge challenge for human-machine interaction. ASIMO is made to be intelligent and as human-like as possible. Designing the machine to reconstruct gestures is an inevitably complex process. ASIMO's gestural performance is a key to synthesis of humans and machines.

In 2007, Toyota released a violin-playing robot (Figure 17.5). Standing at just about 5 feet (152 cm), this robot was also designed to help people with domestic duties. As Toyota describes, their robot was "able to achieve dexterous, human-like hand movement and arm strength when playing the violin," and, "able to achieve vibrato on a violin similar to that created by humans."

Fig. 17.5: The violin-playing robot designed by Toyota.

The fact that robots are approaching realistic human behavior may seem frightening. The ability to realistically gesutrilaze is the primary obstacle for robotic integration into more everyday life. We have robots which automate processes and assemblies, which involve physical gestures. However, intelligent robots need to be able to interpret and interact with the world around them. As gestures can be communicative, they may be the key to free will in robots. If robots are able to master gestural conduct, they will able to use gestures to communicate and learn instinctively like humans. This should not be a fear for humanity. Rather, we should regard it as a large step in our own understanding of the importance of gestures.

18

Neuroscience and Gestures

Summary. When we discuss the human aspects of gestures, we return to the most unique object in nature: the human brain. As with all parts of our experience, the brain is involved with producing gestures. They represent a combination of higher-order thinking and motor function (i.e. movement).

While the entire brain can be studied with reference to gestures, there is a particular type of neurons that we will address here. *Mirror neurons* are neurons that are active both when we perform a behavior and when we see the behavior performed by others. The existence of mirror neurons suggests that gestures represent a fundamental way of learning. We end the chapter with an experiment quantifying the gestures of dancers using motion-sensitive devices to generate music.

18.1 Vilayanur S. Ramachandran and Merlin Donald

Fig. 18.1: Vilayanur S. Ramachandran (1951-)

Fig. 18.2: Merlin Donald (1939-)

© Springer International Publishing AG 2016
G. Mazzola et al., *All About Music*, Computational Music Science,
DOI 10.1007/978-3-319-47334-5_18

Neurologist Vilayanur Ramachandran (Figure 18.1) suggests that mirror neurons have some sort of resonance with our body and gestures. He believes that they are for psychology what replication of DNA is for biology.

In fact, Ramanchandran tested this theory in a series of case studies that revolutionized medical treatment for amputees. Sometimes when a limb is amputated, patients report that they feel pain—in the limb that no longer exists! These medical phenomena, referred to as *phantom limbs*, were first documented in Western culture during the American Civil War. They have been a persistent problem plaguing many amputees. In an elegantly simple design, Ramachandran was the first to propose a way to dispel phantom limbs using only a box and a mirror (see Figure 18.3).

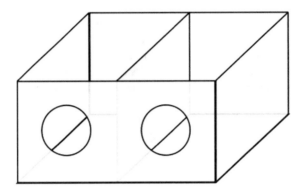

Fig. 18.3: The design for Ramachandran's mirror box. The patient places their amputated arm on one side, and their intact arm in the other, facing the mirror. The visual feedback of gestures made by the arm and hand during mirror therapy appears to activate neuroplasticity resulting in the disappearance of the phantom limb.

In his mirror box experiments, Ramachandran asked participants to insert "both" of their limbs in the box and perform some basic gestural activities (e.g. balling the fist, waving). According to Ramachandran, seeing the limb in the mirror activates the remaining connections in the motor cortex. Over time, the brain is able to receive the proper feedback and realize that the phantom limb does not truly exist. Once this change occurred, patients reported that their phantom limbs suddenly disappeared. This highly gestural approach to treatment relies on mirror neurons and the importance of gestures.

Anthropologist Merlin Donald (Figure 18.2) suggested more broadly that culture evolves through gestures. Specifically, human culture evolves through gestural mimesis, the imitation of gestures. Combining the ideas of Ramachandran and Donald, we propose that there is an important link between neuroscience and gestures.

Gestural imitation is not only a way of learning, but also a way of evolution. However, evolution is not observed until it has already happened. Think, for example, of a culture whose history changed from oral to written tradition. As the practice of handwriting and creating graphic images became of value, written logs of important stories ensued.

We have evolved in parallels based on gestural evolution. We went from hitting stones together, to laying stones in foundations, to colliding particles together. The proliferation of gestures through time is not something viewed only on a short-term timeline, but it can also explains more long-term developments.

18.2 Mirror Neurons, Speech, and Hand Gestures

As mentioned in Chapter 3, the sensorial and motor cortices of the brain contain maps of the human body. These maps can be visually drawn out in the form of a *homunculus* (Figure 18.4). It is a small representation of a human body, whose parts are represented with different proportions according to their sensorial or motoric role.

However, physical movement alone does not adequately explain the phenomenon of gestures, or their role in creation. Cognition and awareness must also be present in order to be creative. Gestures act as a bridge between our cognitive and physical realities. They are the link between the electrical neural network and cognition, the will, and motivation. Gestures are not just a byproduct. It may be best to think of them as the foundation. Without gestures, our neurons would just fire aimlessly and our consciousness would be limited. Without gestures, there is no expression, and ultimately no art.

Without gestures, music would be a mistake!

According to a study done by Shepard and Cooper [12], the time needed to recognize that an object in a picture has been rotated is the time needed to make that rotation in space. This supports the idea that visual, "imaginary" content gets translated into physical movements via gestures. Even our mental capacities are very much linked to gestures. In our minds, we have to make imaginary gestures to understand real spatial transformations. And on the contrary, to understand imaginary rotations, we have to visualize the real gesture that must have taken place.

18.3 The Neuroscience of Imagination

Perhaps one of the most amazing feats of the brain is its ability to re-create complex realities, devise alternate choices and their consequences, and practice gestural actions without physically doing them. All of these things are a part of imagination, a process which allows us to construct, consider, and practice alternate realities without immediately affecting the one we live in. Music is the most important alternative reality in which we find ourselves.

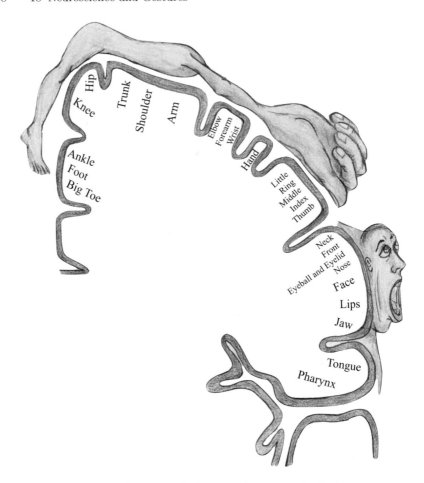

Fig. 18.4: A map of the homunculus across the brain.

Research in the neuroscience of imagination shows that the same areas of the brain are active when someone imagines something as when they perform the action physically. Imagination is therefore a gestural process that doesn't quite reach its full realization.

Imagination not only activates the same parts of the brain as action does, it improves these actions. In 1995, neuroscientist Alvaro Pascaul-Leone conducted an experiment in which subjects were split into two groups (imagination practice and physical practice) and asked to practice a five-finger piano exercise. After several practice sessions, the participants were brought in to *physically* play the exercise. Both groups showed changes in the brain maps of their fingers, and both were able to play better. Initially, the physical practice group performed better, but the imagination practice group caught up to them within a few hours.

So, if you're a musician on the way to a lesson that you didn't practice for, it may benefit you to *imagine* your fingers playing through the piece. Or if you're having trouble throwing that crumpled-up assignment into the trash can, you can *imagine* yourself throwing it. While these examples are entertaining, it is important to keep in mind that actual improvement through imagination requires a large time commitment. Imagination and gestures seem to be inescapably intertwined: we can't have one without the other. We can exploit this fascinating relationship to our benefit and learn to strengthen one using the other and vice versa.

18.4 Mazzola's Gestural Dancing Project: "Dancing the Violent Body of Sound"

In a 2009 project at the University of Minnesota, Guerino Mazzola and the dance professor Rachmi Diyah Larasati discovered that traditional dance in Indonesia from east Java is a natural way to gesturalize Fourier. The traditional rotating dancing style of the dancers (Figure 18.5) represents the superposition of sinusoidal functions in Fourier's theorem, bridging the gap between mathematical theory and art. We can recall the reference to Figure 3.1 for the origin of sinusoidal functions from rotation.

Beyond Larasati and Mazzola, the project was also led by mathematics professor Bill Messing and music student Schuyler Tsuda.

Fig. 18.5: A still from a clip of the traditional Indonesian dance.

According to Larasati, "it is not about narrative but it is about a methodology of communication."

The project presents the use of dance as an expression of Fourier's theory. Fourier's sinusoidal functions were realized through human bodies as the dancers rotated around their axes (Figure 18.6). This models the sinusoidal function introduced in Chapter 3.1.

The next step was to create music from the dancer's movements. Using motion-sensitive Arduino controllers, the project leaders calculated rotation

Fig. 18.6: The University of Minnesota final production.

speeds to replicate the creation of sinusoidal functions. There were sensors measuring acceleration and tilt (Arduino devices, see Figure 18.7). Tilting the wrists upward in motion made the sound go up in volume. The dancers control a segment of their sound in real time. Their movements caused different overtones to jump out. The sound itself came from a collection of recorded overtones of a cello, filtered by the computer according to the dancers' movements.

This direct gesture-to-music relationship creates a literal translation of human movement into sound. This effect was especially prominent because sound waveforms are made from functions based upon the radius of a circle, the basic motion in the dance performed. Through this technology, the dancers were essentially instruments. In general dance follows music, but here music followed dance.

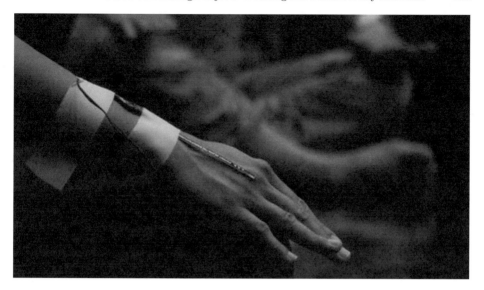

Fig. 18.7: Detail of Arduino sensors on the dancers' body to track their movement, to collect data to be converted it into music.

19

Mathematical Gesture Theory

Summary. What is a *gesture*? This is a concept which everyone knows, but no one is able to define. In that way, it is similar to its counterpart, *time*.

The concept of gesture is very important in art, philosophy, and communication. Formally, a *gesture* is a system of continuous curves connecting points in space and time. A curve that connects a gesture to another one is a hypergesture. We explore the creativity of new music which uses the gesture as a starting point for composition.

The most famous gesture, 'the' gesture, is probably the one of God's Creation of Adam, in the Michelangelo's well-known fresco in Cappella Sistina, shown in Fig. 19.1.

Fig. 19.1: Michelangelo's representation of the Divine Gesture from God to Adam.

The importance of gestures in music is crucial: performers learn specific gestures to play musical instruments, and variations of these gestures allow

© Springer International Publishing AG 2016
G. Mazzola et al., *All About Music*, Computational Music Science,
DOI 10.1007/978-3-319-47334-5_19

different musical-acoustic results. These are the changes of loudness, tempo, and pitch. The importance of gestures can also be gleaned from their historical background, as we will see in the following section.

Fig. 19.2 shows an example of an inappropriate use of gestures to change a musical performance. We are not referring to such cases.

Fig. 19.2: An example of an inappropriate use of gestures to influence and change a musical performance. From the genius of Goscinny and Uderzo.

Western musical notation developed from *neumes*, a system which fixes the shape of the melody and the movements of the choir conductor's hands (chironomic notation) to paper. It is interesting to note that, in a culturally and geographically different environment, musical notation had a similar gestural origin. We have already explored the case of Gregorian music for Western culture. For the Eastern cultures, early Chinese music notation was based on gestures. The notation represented gestural instructions (production) for the player. These were largely created for an instrument called the chin (古琴).

19.1 Historical Roots

Summary. In this section, we will discuss the thoughts of philosophers who emphasized the importance of gestures.

$$-\,\text{\clef{treble}}\,-$$

19.1.1 Tommaso Campanella

Tommaso Campanella (1568-1639) was a Dominican friar and theologian who relied heavily on the importance of the senses in his writings. From his text, whose title can be roughly translated to rational philosophy, he discussed five parts: *dialectica, grammatica, rhetorica, poetica, and historiographia* [9, 63]. His primary message as it pertains to creative gestures is as follows:

...omnes propositiones per singulares tamquam ad digitum exponuntur.
...pointing with your finger is the only certitude.

Campanella simply declares that gestures are the only thing of which we can be completely sure. Gestures do not rely on a byproduct to prove their existence, for a rotation is just a rotation. There may be strokes of paint, but that is only a visible manifestation of the gesture which had to have occurred. Campanella's words speak to the truth that gestures are universal.

19.1.2 Hugues de Saint Victor

Hugues de Saint Victor (c. 1096-1141) was a Saxon theologian. He gives a most adequate definition of a gesture. He describes it as something which is multi-faceted and has factors which must be realized across different realities [62].

Gestus est motus et figuratio membrorum corporis,
ad omnem agendi et habendi modum.
Gesture is the movement and figuration of the body's limbs with an aim,
but also according to the measure and modality
proper to the achievement of all action and attitude.

Hugues de Saint Victor declared that gestures have not only motion, but also an attitude. This is essential to the importance of gestures as a concept. Although gestural motion can be reduced to a mathematical formula, the formula will abandon key information which the original gesture held innately. This means that to fully understand artists and their decisions, we must interpret their works at the gestural level.

19.1.3 Paul Valéry

Paul Valéry (1871-1945) was a French poet and philosopher. He was nominated for the Nobel Prize in Literature on 12 separate occasions. In addition to his artistic expertise, he fostered a special interest in optics. His words on art translate well to mathematics and science.

C'est l'execution du poème qui est le poème.
It's the rendition of a poem which is the poem.
(The interest of science lies in the art of *doing* science.)

Valéry's philosophy was integral to his pursuits in science. However, he defined a separation between an artistic product (such as poem, painting, or song) and the actual artistic making. To him, the art was not the physical byproduct. The art was in the doing and creating, the gesture. He brought this approach to his scientific studies. He believed that the joy of science was not only in the answers that were found, but even more in the questions that were asked.

19.1.4 Jean Cavaillès

Jean Cavaillès (1903-1944) was a French philosopher and mathematician who was particularly interested in the philosophy of science. His words surrounding the role of gestures [10] have had an incredible impact in the direction of this book.

Comprendre est attraper le geste et pouvoir continuer.
Understanding is catching the gesture and being capable of continuing.

Cavaillès' words are the entire foundation of this topic of gestures; they apply to every aspect of the arts. Imagine that you are an improvising musician in a jazz combo. To play something which involves skill would be to play based upon the context which you receive from your fellow musicians. Of course, there is still room for personal flair. However, it only makes sense that you would generally play in the context which the environment decided.

Imagine that you are a solo painter. To convey a coherent message through your work, your gestures must be linked some how. Even Jackson Pollock's gestures have visible connections between them. Just as an orchestra follows the gestures of their conductor, so does an audience interpret the gestures of a painting.

19.1.5 Maurice Merleau-Ponty

The French philosopher Maurice Merleau-Ponty (1908-1961) beautifully summarized the importance of gesture within communication. His illustrated the symbiotic relationship between the communication of ideas and gestures [50].

La parole est un vèritable geste et elle contient son sens comme le geste contient le sien. C'est ce qui rend possible la communication.
Language is a veritable gesture and it contains its sense much as the gesture contains its own. This is what makes communication possible.

Merleau-Ponty combined the ideas of Cavaillès and de Saint-Victor. He proposed that a gesture contains information within itself. Connecting these gestures is like stringing together the words of a sentence. Therefore, we are able to create deeper meaning when we link more gestures together. This brings us to the topic of defining gestures and hypergestures.

19.2 Definition of a Gesture

Besides the fact that the use and the concept of gesture is particularly relevant to music, its precise definition is very recent. The mathematical definition of gesture, given by Mazzola [44], is the following. Let's think of a dancer. He or she touches the stage in discrete points, while moving continuously. If we compare the points to the musical notes, and the continuous curves to the gestures, we can understand their importance for musical performance! And also for composition, because composers also think of final gestures while choosing and writing specific symbols on the musical staff.

Here you are. Let's follow the graphic representation in Fig. 19.3. To the left, we have an abstract structure (music? painting? cooking? nothing yet) of points and connecting arrows, called *skeleton*.[1] To the right, we have a space with the keyboard of the piano indicating the change of position of the hand along this direction to pick a specific pitch, the position above the keyboard (for the 'states' pressed-not pressed key), and the time, the onset. The abstract skeleton is mapped into a system of continuous curves connecting points in space and time, the gesture's *body*. It is clear that we can define gesture for other situations while changing the names of the axis for the space on the right. The number of points and arrows, of course, must be the same. A such mapping is a *gesture*.

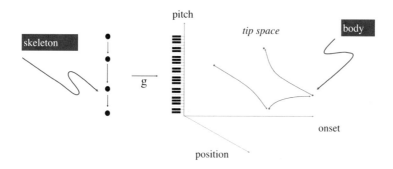

Fig. 19.3: Illustration of the mathematical definition of a gesture, as a mapping from an abstract structure, a skeleton of points and arrows, to a system of continuous curves in space and time, with the same number of points and connecting arrows. The example shown is about the pianistic gesture.

The transformation from notes to sounds is described by the mathematical performance theory of notes [41], from the written information inside the score to the reality of sounds of acoustics. The transformation from gestures 'suggested' by the score into the real movements made by performers, is object of ongoing research, and we already have the first results [36].

How can we get gestures from the score? Let's consider the case of a pianist. If we look literally at the content of a score, we get a series of instructions such as 'play this note at this time with this loudness.' This is more or less what happens for a MIDI instrument! And the change of position should

---

[1] Points and arrows are the starting spot of the mathematical theory of categories, that has been applied to music by Mazzola [41].

happen *instantaneously*. Not even the world's fastest pianist can play with an infinite speed, because it is not allowed by the laws of physics. Sorry for that.

What happens in reality is that the performers need a *finite* time to make a movement. We call symbolic gestures the curve systems of the first type; physical gestures those of the second one. The surface that transforms a gesture of the first type into one of the second is called a world-sheet, see also the blue-red surface in the middle of Figure 19.5. The name and the way to obtain it (via math-theoretical physical methods) comes from string theory in theoretical physics. Again, another interdisciplinary parallelism: elementary particles in physics are no longer seen as 'points' but as vibrating strings; in a similar way, music can be described not only in term of 'points' but also in terms of gestures. This world-sheet surface is also a hypergesture. We will define it in Section 19.3.

The shape of the surface depends on the choice of the force field that shapes the surface. We will not deal with mathematical details here, but we can say that the force field plays the role of the operators that transform the notes of a score into the sounds [41, 36]. Moreover, the time flowing inside the score and the time of the physical performance are qualitatively different: they represent a simplification of time of mental reality versus physical time, see also Section 16.3 for a mathematical description of this approach.

19.3 Hypergestures

In her book about piano performance, Renate Wieland wrote: "Die Klang-berührung ist das Ziel der zusammenfassenden Geste, der Anschlag ist sozusagen die Geste in der Geste." (The sound contact is the target of the embracing gesture, the touch is so to speak the gesture within the gesture). It refers to a more complex concept of gestural utterances. We can make this idea precise in the following sense.

A gesture connects points in space and time. If a gesture is seen as a point itself, the gesture that connects such gesture-points is called *hypergesture*. A hypergesture is a gesture of gestures. Figure 19.4 shows a surface obtained as a loop moving within space along a circle-path.

The concept of hypergesture is very powerful. It helps create incredibly complicated structures of *nested* gestures. An entire symphony, for example, can be studied in terms of hypergestures. With a hierarchical structure, we can describe from top to down the gestures inside an symphonic orchestra. We can think of all the instrumental sections, all the performers of each section, all the micro-movements of the left hand of each violinist This is also a hypergesture. The gestures of each level are composed of smaller gestures of next level.

19.4 A Gesture Suite for Piano

As a *theme* for a musical composition can be built from a melody, a rhythm, a chord, and a timbre. But we can also use a gesture! That's the idea at the base of

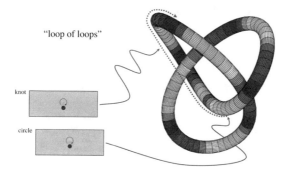

Fig. 19.4: A surface obtained as a loop of loops to represent the concept of hypergesture.

Fig. 19.5: The surface transforming the symbolic gesture into the physical gesture of the musical performance is a hypergesture. From left to right, we have the description of a piano performance; from right to left, the one of the composition from improvisation. Drawings by Maria Mannone.

a recent piano solo suite, *Three Musical Gestures*, composed by Maria Mannone. It is composed of three movements , with variations. The theme is given here by the simple movement up-down in the piano keyboard, the primitive gesture for piano, the one the formalism of world-sheet has been applied to for the first time. The first movement of the suite is *Staccato leggero* (Fig. 19.6), the second *Legato* (Fig. 19.7), and the third *Staccato violento* (Fig. 19.8), with different articulations.[2]

We can think of isolated gestures—for example a single chord. However, we can also combine such isolated gestures to create a more melodic theme.

---

[2] A *movement* inside a musical form always refers to a category of gestures.

We describe such a situation as a collection of circles (isolated gestures) along a hypergestural arrow.[3]

Fig. 19.6: Beginning of the first movement (*Staccato leggero*) of the piano solo suite *Three Musical Gestures*, where the theme is the simple up-down on a fingertip on the keyboard of the piano.

---

[3] It suggests a synthesis of Asian and Western musical traditions: moving a meditative gesture in a dynamic direction.

Fig. 19.7: *Ibidem*, beginning of the second movement, *Legato*.

Fig. 19.8: *Ibidem*, beginning of the third movement, *Staccato violento*.

20

Creativity Theory

Summary. We have seen some examples of both analytical and creative applications of musical gesture theory. Here, we will more systematically describe an approach to musical creativity based on this concept, including an application to improvisation from variations of the Beatles' "Yesterday".

20.1 Defining Creativity

Creativity has been explored by many scholars. We seek to define the process of creativity as something that can be universally attained. When we create there is a general process which we propose is universally followed by all artists. Notably, this process is reversed for consumers and critics of art.

Artists start with gestures. The stroke of a pen or paintbrush, a pirouette, the wiggling of fingers... All these are gestures. They're not primitive, nor are they mindless.

Gestures are then fit into some sort of process. Perhaps, a pianist chooses to follow a standard blues chord progression. Perhaps, they run a broken rake across their keyboard. Whether or not an artist is anti-establishment makes no difference. The fact is that when there are patterns that can exist, there are also instances in which there seem to be no patterns. Both are possible cases which the artist chooses to do over the span of a series of gestures.

By the end of it all, we reach facts. This level of factual information is comparable to the neutral niveau of Molino's tripartion [53]. It is from this point that one may begin tracing and trying to understand the creative process. An interpreter will only experience the facts (Figure 20.1). It is their mission to try to understand the processes and gestures and how they constitute the final product and message.

© Springer International Publishing AG 2016

G. Mazzola et al., *All About Music*, Computational Music Science,

DOI 10.1007/978-3-319-47334-5_20

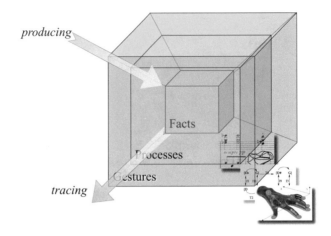

Fig. 20.1: A diagram that outlines the creative process and its reverse.

To be creative is a question of its own right. We propose that there are seven types of purposes behind the act of creating something. Typically, they are followed in this order. Examples are given below each step.

1. *Exhibit an open question.*
   What is life? Why is this minority group being repressed? How do I feel?
2. *Define its context.*
   Children, adults, clones. Africa, Europe, Asia. Antiquity, Renaissance, post-World War II. Where are we and who is around us?
3. *Find a critical concept.*
   Morals, values, social ideals.
4. *Exhibit its "walls".*
   What are the social constructs which limit behavior? How is philosophy and reasoning inhibiting creativity? How could creativity destroy civilization?
5. *Soften the walls.*
   Be the devil's advocate. Provide alternatives. Propose the opposite.
6. *Extend them.*
   Make a new reality. Suppose that the opposite is true.
7. *Evaluate the extended concept.*
   Is this right? Is this wrong? Should we change or adapt to this new reality? Is it even possible?

20.1.1 Example of Creative Processes in Beethoven's op. 109, Third Movement

According to William Kinderman [28], it is essential that studying the creative process of an individual requires some analysis of the *entire* evolution of their processes. This means that it is important to look at something through all of its evolutions. Here is Kinderman's statement:

> A microscopic scrutiny of sketches according to a particular analytical technique is usually too limited to yield much insight into the creative process; a more promising approach would give attention to the relationships between successive versions, since the comparison sometimes reveals a network of compositional changes.

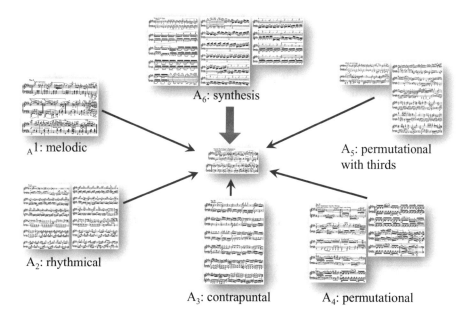

Fig. 20.2: The culmination of the various variations in Beethoven's Sonata op. 109/3, pointing to the theme as the primary source, but also the lifeline.

In terms of creativity, the sixth creative movement is a creative target issued from the characteristic wall extensions of the first five variations of the movement's critical theme. Beethoven's approach is intensive. He looks into the 'anatomy' of the theme and searches for its substance (Figure 20.2). The six variations are not independent of the theme. They are directed towards the theme. Each variation stresses a particular aspect of the theme. In this way, we can see that Beethoven was not only interested in establishing an open question and providing context (the theme) but he exhibited, softened and extended the walls with each variation.

20.2 The Creative Strategy in "¡Ornette!" from the Movie *Imaginary Time*

Free jazz is not just a bunch of musicians standing together and playing random notes on their instruments. There is a communication to free jazz, as there is to all music. Mazzola has demonstrated this to us through his 2014 performance at the Airegin Jazz club in Yokohama, Japan. The result was an 85-minute free jazz feature with Mazzola on piano, drummers Heinz Geisser and Shiro Onuma, a Japanese audio-visual team, and a Swiss post-production team. Between Christmas night and the following day, the group gathered with microphones and cameras tracking every space within the cozy club.

A pamphlet posted with the online release of the recording describes the artists' intent:

> In *Imaginary Time*, what you hear is what you see, a deep connectivity of musical and gestural expression, exquisitely recorded by a Japanese team of audio and video technicians, and edited by a first class music film professional in Europe. To a great extent, the ease of communicating this music is grounded in gestural language that would be difficult to grasp on an exclusively audio channel. Its beauty can be experienced as a flourishing dance of sound, a deployment of the imaginary time of artistic consciousness. This music is quite the opposite of blind spontaneous trial and error, it is grounded in the persistent accumulation of a life's richness of images, movements and miracles.

> Our product is the faithful expression of our intentions, down to the least details of audio and video shaping. We have been very attentive to the diligent changes of perspectives between long shot and split screen views, to make evident the interplay in the musical drama.

In their setlist, they performed a creative recomposition of the Beatles' "Yesterday". Motifs can be heard with altered rhythms and pitches. Eventually, one may feel as if the piece has entirely drifted from the original content. However, the connection can be seen in the artists' intent and gestural motion. Titled, "¡Ornette!", the piece is a tribute to the great free jazz saxophonist, Ornette Coleman. He pioneered the ever-growing avant-garde stage of jazz. His music featured dramatic and complex shifts in time. Between Mazzola and the two drummers at the Airegin, a whole conversation of time was created.

Part VI

References, Index

References

1. Adorno Th W: Zu einer Theorie der musikalischen Reproduktion. Suhrkamp, Frankfurt/M 1946
2. Agustín-Aquino O A, J Junod, G Mazzola: Computational Counterpoint Worlds. Springer, Heidelberg 2015
3. Barthes R: Éléments de sémiologie. Communications 4/1964
4. Békésy G von: Experiments in Hearing. McGraw-Hill, New York 1960.
5. Boissière A: Geste, interprétation, invention selon Pierre Boulez. Revue DEMéter, déc. 2002, Univ. Lille-3
6. Braxton A: Composition No. 76, 1977 http://tricentricfoundation.org
7. Braxton A: For Four Orchestras, Composition No. 82. LP Album. 1978
8. Cage J: 4'33". Edition Peters 1952
9. Campanella T: Philosophiae rationalis partes quinque. Paris, Du Bray. 1637-38.
10. Cavaillès J: Méthode axiomatique et formalisme, Hermann 1938
11. Châtelet G: Figuring Space. Kluwer 2000
12. Shepard R N and L A Cooper: Mental Images and Their Transformations. MIT Press 1982
13. Darwin Ch: The expression of the emotions of man and animals. John Murray, London, England 1872
14. Descartes R: Compendium musicae. 1618
15. Euler L: De harmoniae veris principiis per speculum musicum representatis (1773). In: Opera Omnia, Ser. III, Vol. 1 (Ed. Bernoulli, E et al.). Teubner, Stuttgart 1926
16. Fletcher N H and Th D Rossing: The Physics of Musical Instruments. Springer 2010
17. Frova A: Fisica nella musica. Zanichelli, Bologna 1999
18. Fux J J: Gradus ad Parnassum (1725). Dt. und kommentiert von L. Mitzler, Leipzig 1742; English edition: The Study of Counterpoint. Translated and edited by A Mann. Norton & Company, New York, London 1971
19. Geisser H, G Mazzola, S Onuma: Imaginary Time. DVD, to be released 2016
20. Gabrielsson A: Expressive Intention and Performance. In: Steinberg R (ed.): Music and the Mind Machine. Springer, Berlin et al. 1995
21. Goupillaud P, A Grossman, and J Morlet. Cycle-Octave and Related Transforms in Seismic Signal Analysis. Geoexploration, 23:85-102, 1984

© Springer International Publishing AG 2016
G. Mazzola et al., *All About Music*, Computational Music Science,
DOI 10.1007/978-3-319-47334-5

180   References

22. Hanslick E: Vom Musikalisch-Schönen. Breitkopf und Härtel (1854), Wiesbaden 1980
23. Hatten R: Interpreting Musical Gestures, Topics, and Tropes. Indiana University Press 2004
24. Hawking S: A Brief History of Time. Bantham Books, New York 1988
25. Hebb D O: The Organization of Behavior. Wiley & Sons, New York 1949
26. Hesse H: Das Glasperlenspiel (1943). Suhrkamp, Frankfurt/M. 1973
27. Jakobson R: Linguistics and Poetics. In: Seboek, TA (ed.): Style in Language. Wiley, New York 1960
28. Kinderman W: Artaria 195. University of Illinois Press, Urbana and Chicago 2003
29. International Conference on New Interfaces for Musical Expression. http://www.nime.org
30. Koelsch S et al.: Children Processing Music: Electric Brain Responses Reveal Musical Competence and Gender Differences. Journal of Cognitive Neuroscience Vol. 15(5), 2003
31. Leman M and R I Godøy (eds.): Musical Gestures. Routledge, New York and London 2010
32. Lewin D: Generalized Musical Intervals and Transformations (1987). Cambridge University Press 1987
33. Lewin D: Musical Form and Transformation: 4 Analytic Essays. Yale University Press, New Haven and London 1993
34. Langer S: Feeling and Form, Routledge and Kegan Paul, London 1953
35. Lazarus, R S: Emotion and Adaptation. Oxford University Press, New York 1991
36. Mannone M and G Mazzola: Hypergestures in Complex Time: Creative Performance Between Symbolic and Physical Reality. Springer proceedings of the MCM15 Conference, 2015
37. Mazzola G et al.: A Symmetry-Oriented Mathematical Model of Classical Counterpoint and Related Neurophysiological Investigations by Depth-EEG. In: Hargittai I (ed.): Symmetry II, CAMWA, Pergamon, New York 1989
38. Mazzola G and E Hunziker: Ansichten eines Hirns. Birkhäuser, Basel 1990
39. Mazzola G: *Synthesis*. SToA 1001.90, Zürich 1990
40. Mazzola G: Semiotic Aspects of Musicology: Semiotics of Music. In: Posner R et al. (Eds.): A Handbook on the Sign-Theoretic Foundations of Nature and Culture. Walter de Gruyter, Berlin and New York 1998
41. Mazzola G: The Topos of Music. Birkhäuser, Basel 2002
42. Mazzola G et al.: The Topos of Music III: Gesture. (Third volume of second edition of [41]) Springer, Heidelberg 2017
43. Mazzola G: La vérité du beau dans la musique. Delatour, Paris 2007
44. Mazzola G and M Andreatta: Diagrams, Gestures, and Formulas in Music. Journal of Mathematics and Music 2007, Vol. 1, no. 1, 2007
45. Mazzola G and P B Cherlin: Flow, Gesture, and Spaces in Free Jazz—Towards a Theory of Collaboration. Springer Series Computational Music Science, Heidelberg 2009
46. Mazzola G: Musical Performance—A Comprehensive Approach: Theory, Analytical Tools, and Case Studies. Springer Series Computational Music Science, Heidelberg, December 2010
47. Mazzola G, J Park, F Thalmann: Musical Creativity. Springer, Heidelberg 2011

48. Mazzola G: Musica e Matematica: due movement aggiunti tra formula e gesti. In: Bertocci C and P Odifreddi (eds.): La Matematica: Suoni, Forme, Parole, Einaudi, 2011, pp. 159-198]

49. Society for Mathematics and Computation in Music. http://www.smcm-net.info

50. Merleau-Ponty M: Phénoménologie de la perception. Gallimard 1945

51. Milmeister G: The Rubato Composer Music Software: Component-Based Implementation of a Functorial Concept Architecture. Springer Series Computational Music Science, Heidelberg 2009

52. Mozart W A: Walzer und Schleifer mit zwei Würfeln zu componieren ohne Musikalisch zu seyn, noch von der Composition etwas zu verstehen. Schott, Mainz 1984

53. Nattiez J-J: Fondements d'une Sémiologie de la Musique. Edition 10/18 Paris 1975

54. Oechslin M S et al.: Hippocampal volume predicts fluid intelligence in musically trained people. Hippocampus, 23(7), 552-558, 2013

55. Pascual-Leone A et al.: Modulation of muscle responses evoked by transcranial magnetic stimulation during the acquisition of new fine motor skills. Journal of Neurophysiology, 74(3), 1037-1045, 1995

56. Penrose R: The Road to Reality. Vintage, London 2002

57. Petsche H et al.: EEG and Thinking: Power and Coherence Analysis of Cognitive Processes. Austrian Academy of Sciences, 1998

58. Pierce A: Deepening Musical Performance through Movement. Indiana University Press 2007

59. Rahaim M: Musicking Bodies. Wesleyan University Press, Middletown CT 2012

60. Riemann H: Vereinfachte Harmonielehre oder die Lehre von den tonalen Funktionen der Akkorde. London 1893

61. Russell J A: A Circumplex Model of Affect. J. of Personality and Social Psychology, Vol. 39, pp. 1161-1178, 1980

62. Schmitt J-C: La raison des gestes dans l'Occident médiéval. Gallimard, Paris 1990

63. Schmidt-Biggemann W: Topica universalis. Meiner, Hamburg 1983

64. Slater E and A Meyer: A contribution to a pathography of the musicians: 1. Robert Schumann. Confinia Psychiastricia, 2: 65-94, 1959

65. Juslin P N and J A Sloboda: Handbook of Music and Emotion (Affective Science). Oxford University Press, 2010

66. Tomkins S S: Affect, imagery, consciousness (Vol. 1). Springer, Oxford, England 1962

67. Wieland R and J Uhde: Forschendes Üben. Bärenreiter-Verlag, Kassel, Germany 2002

68. Wieser H-G and G Mazzola: Musical consonances and dissonances: Are they distinguished independently by the right and left hippocampi? Neuropsychologia Vol. 24 (6), pp. 805-812, 1986

69. Wieser H-G and G Mazzola: EEG responses to music in limbic and auditory cortices. In: Engel J Jr, Ojemann G A, Lüders H O, Williamson P D (eds.): Fundamental mechanisms of human function. Raven, New York 1987

70. Zarlino G: Le istitutioni harmoniche (1573). Facsimile edition: Gregg Press 1966

Index

© Springer International Publishing AG 2016
G. Mazzola et al., *All About Music*, Computational Music Science,
DOI 10.1007/978-3-319-47334-5

Printed in the United States
By Bookmasters